## 旅酒マップ　TABI-SAKE MAP

# はじめに

旅をしなければ、飲めない酒がある。
探さなければ、見つからない酒がある。
旅が、宝探しになる。
日本人の「心からのおもてなし」に出会える場所に
旅酒はあります。

その地を訪れる理由、それは、
そこに行かなければ飲めない酒があるから。

旅酒は、日本各地の魅力的な地域限定で販売されている
“旅先でしか買えないお酒”です。

情報社会、物流の発展により、日本中どこに行っても
同じようなサービスを受けられるようになってきたこの時代に、
あえてその地でしか買えないお酒を造りました。

「旅が宝探しになるように！」

旅酒を探しに旅に出る。
その旅のお供にしていただくために、
この旅酒ガイドブックを出版しました。

この本を片手に、
旅酒を探す旅に出てみてはいかがでしょうか。

# Preface

There is SAKE you must meet.
There is SAKE you must look for.
A trip becomes a treasure hunt.
Find your TABI-SAKE
in the heart of OMOTENASHI,
the spirit of Japanese hospitality.

Why are you going there?
That's because there is *sake* there that you cannot drink anywhere else.

TABI-SAKE is a limited edition product that you can purchase only on a trip.
If you travel to attractive places in Japan, you will surely find it.

Nowadays,
with the development of information technology and logistics,
you can receive the same services anywhere in Japan.
Still, we have dared to make special *sake*
that you cannot purchase unless you visit the place.

**Make your trip a treasure hunt.**

Take a trip to find TABI-SAKE.
We have published this TABI-SAKE guidebook
to serve you as you travel.

With this book in your hand,
why not set off to find TABI-SAKE?

◎日本酒の種類　kinds of Sake (Japanese liquor brewed by adding water and malt to rice)

| 名称<br>Name | 使用原料<br>Materials used | 精米歩合※1<br>Polishing Raito | 旅酒番号<br>TABI-SAKE No. |
|---|---|---|---|
| 吟醸酒<br>Ginjo-shu | 米，米麹，醸造アルコール※2<br>rice, koji, brewers alcohol | 60％以下<br>Up to 60% | — |
| 大吟醸酒<br>Daiginjo-shu | 米，米麹，醸造アルコール<br>rice, koji, brewers alcohol | 50％以下<br>Up to 50% | — |
| 純米吟醸酒<br>Junmai Ginjo-shu | 米，米麹<br>rice, koji | 60％以下<br>Up to 60% | 6, 7, 10, 15, 16, 17, 19, 20, 21, 22, 24, 32, 36, 38 |
| 純米大吟醸酒<br>Junmai Daiginjo-shu | 米，米麹<br>rice, koji | 50％以下<br>Up to 50% | 12, 13, 37, 39, 46 |
| 純米酒<br>Junmai-shu | 米，米麹<br>rice, koji | — | 1, 3, 11, 18, 23, 29, 34, 42, 43, 44, 45, 48, 49 |
| 特別純米酒<br>Special Junmai-shu | 米，米麹<br>rice, koji | 60％以下または特別な製造方法<br>Up to 60% or specially processed | 2, 4, 9, 14, 27, 28, 31, 35, 40, 47, 50 |
| 本醸造酒<br>Honjozo-shu | 米，米麹，醸造アルコール<br>rice, koji, brewers alcohol | 70％以下<br>Up to 70% | — |
| 特別本醸造酒<br>Special Honjozo-shu | 米，米麹，醸造アルコール<br>rice, koji, brewers alcohol | 60％以下または特別な製造方法<br>Up to 60% or specially processed | 51 |

※１　精米歩合：精米工程で米をどれだけ精米したかを，元の玄米の重量に対する白米の重量の割合を示す数字。
Polishing Raito : A figure expresses the percentage of polished rice (in weight) relative to the brown rice (unpolished).

※２　醸造アルコールの使用量は白米の重量の10％以下に制限されている。
There should be no more than 10% brewers alcohol relative to the rice by weight.

◎日本酒以外の酒　other kinds of liquor

| 名称<br>Name | 定義 | Description | 旅酒番号<br>TABI-SAKE No. |
|---|---|---|---|
| 焼酎<br>Shochu | 芋・麦・米・糖蜜類などを原料とした日本の蒸留酒 | Japanese distilled spirits made from various ingredients, including sweet potato, buckwheat, rice and sugar cane | 8, 25, 41 |
| 琉球泡盛<br>Ryukyu Awamori | 米・泡などを原料とした沖縄特産の蒸留酒 | Okinawa's distilled spirits made from rice or foxtail millet | 5, 30 |
| 梅酒<br>Plum Wine | 梅を焼酎などに漬けて作る果実酒 | Japanese liquor made from plum steeped in alcohol (shochu and so on) and sugar | 26, 33 |
| ウイスキー<br>Whisky | 大麦・ライムギなどを原料として作る蒸留酒 | distilled spirits from malted grain, especially barley or rye | 52 |

# 日光 （栃木県）

世界遺産にも登録された日光東照宮をはじめとする歴史的な建造物と豊かな自然美。夏は避暑地、冬は温泉地として、永きにわたり愛されてきた日光。日本国内はもちろん、海外からも多くの旅行客が訪れる日本屈指の観光地です。江戸時代から続く歴史と自然美をゆっくりと感じながら、日光の「おもてなし」を味わい、貴重な時間を過ごしてはいかがでしょうか。

Nikko is fairly well known for the World Heritage Site, the Nikko Toshogu Shrine, and many other historical monuments, all surrounded by the rich natural beauty. As a summer resort during the summer and as a spa during the winter, it has been loved by many people for a long time. Not only a popular tourist resort domestically but also internationally. Why not spend a valuable time at Nikko by slowly feeling the natural beauty?

## 旅 TABI-SAKE 酒

**NIKKO**

気負わず楽しめる純米酒。軽快でなめらかな旨さが料理を次に進ませます。炊きたてのご飯を感じさせるふくよかな香りに、柑橘系の果実を連想させる香りがわずかに加わります。やさしい甘みと爽やかな酸味が調和し、旨みとコクを感じさせる味わいです。

A light and smooth taste makes you want to eat more. A scent of freshly cooked rice and a little bit of citrus scent. A gentle sweetness and a pleasant acid taste are harmonized well, creating a great flavor.

**醸造元：** 第一酒造株式会社
（1673年創業／栃木県）
**種類：** 日本酒 純米酒
**度数：** 14度以上15度未満
**精米歩合：** 65%

**Brewery** : Daiichi Shuzo Co., Ltd.
(Since 1673／Tochigi)
**Kind** : Junmai-shu
**Alcohol Content** : 14%-15%
**Polishing Ratio** : 65%

甘辛度
Sweet-dryness

## 主な行き方 Major Directions

羽田空港＝(20分)⇒浜松町駅＝(4分)⇒東京駅＝(新幹線50分)⇒宇都宮駅＝(50分)⇒日光駅

Haneda Airport＝(20 min)⇒Hamamatsucho Station＝(4 min)⇒Tokyo Station＝(Shinkansen 50 min)⇒Utsunomiya Station＝(50 min)⇒Nikko Station

## 旅の宝 Treasure Hunt

### ▶①日光東照宮

日本を代表する世界遺産「日光の社寺」。そのなかでももっとも有名な「日光東照宮」は徳川家康がまつられた神社で、現在の社殿群は、そのほとんどが寛永13年3代将軍家光による「寛永の大造替」で建て替えられたもの。境内には国宝8棟、重要文化財34棟を含む55棟の建造物が並び、その豪華絢爛な美しさは圧巻です。全国各地から集められた名工により、建物には漆や極彩色がほどこされ、柱などには数多くの彫刻が飾られています。

### ▶②華厳ノ滝

48もの滝が点在する日光周辺で、最も有名ともいえるのが華厳ノ滝。中禅寺湖の水が高さ97mの岸壁を一気に落下する壮大な滝で、自然が作り出す雄大さと華麗な造形美の両方を楽しむことができます。5月には見事な新緑、6月にはたくさんのイワツバメが滝周辺を飛び回り、1月から2月にかけては十二滝と呼ばれる細い小滝が凍るため、滝全体がブルーアイスに彩られ、四季折々に違った景色を堪能することができます。

### ▶①Nikko Toshogu Shrine

The shrines and temples in Nikko represent Japan's World Heritage Site. Out of all the shrines and temples, the most famous one is the Nikko Toshogu Shrine which Ieyasu Tokugawa is enshrined at. Almost all of the current main Shrine group were rebuilt by the 3rd Shogun Iemitsu Tokugawa during the "Great rebuilding of the Kanei era". In the grounds, there are 55 buildings, including 8 buildings of national treasures and 34 buildings of Important Cultural Properties, and the absolute gorgeousness is overwhelming. By master craftsmen that have been collected fromall over the country, lacquer and the use of rich color are decorated in the building, and numerous sculptures are decorated in some pillars.

### ▶②Kegonnotaki Watertall

Out of 48 waterfalls that are scattered around Nikko, Kegonnotaki watertall can be said to be the most famous one. The water of Lake Chuzenji falls straight down from a 97m height quay and you can enjoy both the majesty and the formative art created by the nature. In May, the rich green, in June, the Asian house martin flying around the waterfall, and from January to February, "the Junitaki" which small waterfalls freeze and color the whole waterfall in blue ice; you can enjoy different scenes in each season.

## Brewery's recommended spot
## 蔵元おすすめスポット

### 鬼怒川ライン下り

日光市鬼怒川温泉大原1414
1414, Kinugawaonsenohara, Nikko-shi

大自然が創造した渓谷美の極みとして名高い鬼怒川の名物。

### Going down the Kinugawa River

Kinugawa River is famous for its beautiful valley which the nature created.

### 鬼怒楯岩大吊橋

日光市鬼怒川温泉大原1436
1436, Kinugawaonsenohara, Nikko-shi

鬼怒川温泉街の南部と名勝「楯岩」を結ぶ全長140mの歩道専用吊橋。

### Kinutateiwa Suspension Bridge

The bridge, only available to walk, connects the spa town of Kinugawa and the Tateiwa.

## 旅酒が見つかる場所　Shop

- **鬼怒川観光ホテル 売店**
  日光市鬼怒川温泉滝359-2
  0288-77-1101
- **三本松茶屋**
  日光市中宮祠2493
  0288-55-0287
- **湯沢屋酒店**
  日光市湯西川766
  0288-98-0423
- **おみやげ処 すみ屋**
  日光市鬼怒川温泉大原1396-8
  028-877-0821
- **本家伴久**
  日光市湯西川749
  028-898-0011
- **あさやホテル 売店**
  日光市鬼怒川温泉滝813
  028-877-1111
- **鬼怒川グランドホテル 売店**
  日光市鬼怒川温泉大原1021
  028-877-1313
- **おみやげの店 ながた**
  日光市中宮祠2478
  028-855-0250
- **ローソン 日光東照宮前店**
  日光市本町2-32
  0288-25-6588
- **早見商店**
  日光市松原町12-5
  0288-54-0808

※最新の販売店情報はホームページでご確認ください。
Please check our website for the latest retail shop information.

# 02

# 湯布院 （大分県）

緑豊かな温泉地、湯布院。高原の冷涼な気候でとても過ごしやすい観光地。地場の食材を使用した多彩な食も魅力のひとつ。温泉・食・自然・文化……すべてが揃う湯布院はリピーターも多いといわれています。その魅力を味わうために、一度足を運んでみてはいかがでしょうか。

Yufuin, a spa resort surrounded by nature. The climate of the plateau creates a pleasant tourist resort. The food using a variety of local ingredients is also one of the attraction. Spa, food, nature, and culture... Yufuin, having all the good points, is said to have a lot of repeaters. To enjoy the attractions of Yufuin, why don't you once visit Yufuin?

## 旅酒 TABI-SAKE YUFUIN

力強くふくよかな米の旨みと香り。酒米の王様「山田錦」を九重自慢の伏流水で醸します。酒が自然の恵みであることを感じさせる、米の旨みが印象的な深い飲みごたえ。

A vigorous, and rich rice flavor and scent. The king of rice sake, "Yamadanishiki," is brewed by the underground water of Kokonoe. You can feel that the sake is a blessing of the nature and the Special Junmai-shu is an impressive taste of rice.

**醸造元：**八鹿酒造株式会社（1864年創業／大分県）
**種類：**日本酒 特別純米酒
**度数：**15度
**精米歩合：**60%

**Brewery** : Yatsushika Brewery Co., Ltd.（Since 1864／Oita）
**Kind** : Special Junmai-shu
**Alcohol Content** : 15%
**Polishing Ratio** : 60%

甘辛度
Sweet-dryness
甘 Sweet — 辛 Dry

## 主な行き方 Major Directions

大分空港／バス＝(60分)⇒大分駅＝(50分)⇒湯布院駅
Oita Airport／Bus＝(60 min)⇒Oita Station＝(50 min)⇒Yufuin Station

## 旅の宝 Treasure Hunt

### ▶①金鱗湖

湯布院のシンボルとなっている観光スポット。金鱗湖湖底から温泉が湧いており、水温が高く、冬の早朝には湖面から霧が立ち上り、幻想的な風景が広がります。新緑や紅葉も美しく、四季折々でその姿を楽しむことができる名所です。

### ▶①Lake Kinrin

Lake Kinrin is a symbol of Yufuin tourist resort. Hot spring flows out from the bottom of the lake, so the temperature of the water is high. Thus, you can enjoy a romantic view. The fresh green and red leaves are also beautiful and you can enjoy certain views for each season.

### ▶②由布岳

湯布院の北東部にそびえる由布岳。豊後富士とも称されるこの山は、多くの登山客が訪れる大分を代表する山のひとつ。『古事記』や『豊後国風土記』といった日本古来からの文献にも登場する名峰。岩峰を東西に突き上げるようにして屹立する姿は実に雄大です。

### ▶②Mt. Yufu

Mt. Yufu rises in the north west of Yufuin and is also called the Bungo Fuji. Many mountain climbers visit this mountain which represents one of Oita's mountains. It is also depicted in traditional Japanese literatures such as *Kojiki* and *Bungo Fudoki*. The figure of the pinnacles of the mountain which push up to the west and east is very dynamic.

## Brewery's recommended spot
## 蔵元おすすめスポット

### くじゅう連山(ミヤマキリシマ)

玖珠郡九重町大船山、平治岳、三俣山、星生山、黒岩山、猟師山、一目山、合頭山、泉水山など
玖珠郡九重町
Kokonoe-machi, Kusu-gun

毎年、5月末〜6月初旬頃に、ミヤマキリシマのピンクの花が咲き誇り、山がピンクに染まる。この時期は多くの登山客でにぎわう。

### Kuju-Renzan Massif (Miyamakirishima)

From the end of May to the middle of June every year, the pink flowers of Miyakirishima bloom and the whole mountain becomes pink. During this season, there are many hikers.

## 旅酒が見つかる場所 Shop

- **清美堂 駅前通り店**
  由布市湯布院町川上3727-4
  0977-84-2414
- **下谷商店**
  由布市湯布院町川上3724-10
  0977-84-2057
- **金鱗堂**
  由布市湯布院町川上3048-1
  0977-84-4486
- **花夢遊**
  由布市湯布院町川上1541-11
  0977-85-2717
- **のいちご**
  由布市湯布院町川上1541-11
  0977-85-2717
- **一休商店**
  由布市湯布院町川北4-1
  0977-85-2801
- **九州フードマーケット**
  由布市湯布院町川上1503-3
  0977-28-8615
- **ゆふいん 道の駅**
  由布市湯布院町川北899-76
  0977-76-5607
- **山荘 無量塔**
  由布市湯布院町川上1264-2
  0977-84-5000
- **ゆふいんホテル 秀峰館**
  由布市湯布院町川上2415-2
  0977-84-5111
- **湯布院 山灯**
  由布市湯布院町川上1553
  0977-84-5310
- **湯布院 薫風工房**
  由布市湯布院町川上1511-1
  0977-28-8858
- **大分銘品蔵**
  大分市要町1-40 JR大分駅 豊後にわさき市場内
  097-513-7061

※最新の販売店情報はホームページでご確認ください。
Please check our website for the latest retail shop information.

**◆他にも素敵な場所があります。あなただけの旅の宝を探してみましょう!**
**There are other wonderful places to visit. Let's take a trip to find a treasure of your own!**

## MATSUE・IZUMO -Shimane-

# 松江・出雲 （島根県）

全国の神様が集う神秘の古社、出雲大社をはじめとする日本有数のパワースポット。日本神話ゆかりの地や、歴史的な建物や遺跡を巡る観光ルート、紅葉や絶景といった自然美を楽しむコース。名産の出雲そばを食べ歩くというのもあり。さまざまな魅力が詰め込まれた出雲を堪能してみてはいかがでしょうか。

Gods of all over the country assemble at mysterious old shrines, including the Izumo Taisha Shrine which is one of the locations thought to be flowing with mystical energy in Japan. A tourist route that travels around places that are famous for its associations with Japanese mythology, and historical buildings and ruins. A course of tourism that enjoys natural beauty such as red leaves and superb views. An eating tour of the well-known product Izumo soba. Please enjoy all sorts of attractions of Izumo.

## 旅 TABI-SAKE 酒

### MATSUE ・ IZUMO

島根県産酒造好適米「五百万国」を60%に精米し、低温でゆっくりと醗酵させて米の味を引き出します。ふくよかで濃醇ながら、スッキリとした後切れの良い味わい。

Using 60% of cleaned rice of "Gohyakumangoku," a rice suited for making sake from Shimane prefecture, and fermenting it at a low temperature slowly so that the flavor of rice gets out. The flavor although rich and strong, it is fresh at the end.

**醸造元**：板倉酒造有限会社
（1871年創業／島根県）
**種類**：日本酒 純米酒
**度数**：15%
**精米歩合**：65%

**Brewery** : Itakura Shuzo Co., Ltd.
(Since 1871／Shimane)
**Kind** : Junmai-shu
**Alcohol content** : 15%
**Polishing Ratio** : 65%

**甘辛度**
Sweet-dryness
甘 — 辛
Sweet — Dry

## 主な行き方 Major Directions

出雲空港／バス＝(30分)⇒松江駅＝(40分)⇒出雲市駅
Izumo Airport／Bus＝(30 min)⇒Matsue Station＝(40 min)⇒Izumo City Station

## 旅の宝 Treasure Hunt

### ▶①出雲大社

神々の国であるとされる出雲。なかでも出雲大社は神話の時代からの歴史を持ち、パワースポット、縁結びの神様として年間200万人を超える参拝客でにぎわっています。旧暦の10月には日本各地の神々が出雲に集まるとされ、神迎祭が行われます。足を踏み入れれば、厳かな雰囲気を確かに感じることができます。

### ▶①Izumo Taisha Shrine

Izumo is known for the country of gods. Above all, the Izumo Taisha Shrine where 2 million visitors of shrines and temples flourish annually has a history of thousand years, a location thought to be flowing with mystical energy, and the god of marriage. In October according to the lunar calendar, it is said that the gods all over Japan gather at Izumo and a welcoming festival for gods is held. Once people step in, they can surely enjoy the austere atmosphere.

### ▶②鰐淵寺

うっそうと繁る木々のなかにある鰐淵寺は、重要文化財を多数とりそろえた古寺です。新緑の季節には青々と繁る草木が美しく、紅葉の季節には山全体が赤く染め上がります。紅葉が敷き詰められた参道を歩くのはこの時季の醍醐味。豊かな自然美を感じることができます。

### ▶②Gakuenji Temple

Gakugenji Temple, found inside densely luxuriated trees, is an old Shrine that has many Important Cultural Properties. In the season of new green, the green luxuriated plants are beautiful, and in the season of red leaves, the maples dye the whole mountain in red. Walking the path covered with full of maple leaves is the real pleasure of this season. People can feel and enjoy the rich natural beauty.

**Brewery's recommended spot**
**蔵元おすすめスポット**

## 稲佐の浜

出雲市大社町杵築北稲佐
Kitainasa, Kizuki, Taisha-cho, Izumo-shi

## 日御碕

出雲市大社町日御碕
Hinomisaki, Taisha-cho, Izumo-shi

夕日にまつわるストーリー「日が沈む聖地出雲」が日本遺産に認定された。出雲神話の舞台となった「稲佐の浜」と「日御碕」は、夕日の絶景として知られている。

## Inasanohama Beach and Cope Hinomisaki

The story, "Sunset in the Sacred Land of Izumo," has been registered as the Japanese heritage. The Beach and the Cope are mentioned in Izumo Shinto mythology, and both are famous for sunset view.

**旅酒が見つかる場所** **Shop**

●**アトネスいずも**
出雲市駅北町10-3
0853-21-6783

●**松江フォーゲルパーク**
松江市大垣町52
0852-88-9800

※最新の販売店情報はホームページでご確認ください。
Please check our website for the latest retail shop information.

◆**他にも素敵な場所があります。あなただけの旅の宝を探してみましょう!**
**There are other wonderful places to visit. Let's take a trip to find a treasure of your own!**

04

## MIYAJIMA -Hiroshima-

# 宮島 （広島県）

日本三景のひとつとして名高い宮島。誰もが一度は訪れてみたいあこがれの地であり、日本人の旅の心の原点です。世界文化遺産にも登録された厳島神社、国の重要文化財となっている建造物、先人達が心を奪われた豊かな自然を味わいながら日本古来からの名所を堪能してみてはいかがでしょうか。

Miyajima Island is one of the three great views of Japan. A place where everyone wants to visit and see the view at least once and is the starting point of travel for Japanese people. Itsukushimajinja Shrine registered as one of the World Heritage Site and buildings being the country's Important Cultural Properties. Why don't you enjoy the traditional Japanese sights and the rich nature which our predecessors have been fascinated by?

## 旅 TABI-SAKE 酒

### MIYAJIMA

辛口ながら、お米の旨みを感じる飲み飽きしない食中タイプの純米酒。冷酒からお燗まで幅広く楽しめます。江戸時代の町並みが残る竹原にある蔵で醸造しています。

With a dry taste and having its rice flavor, the Special Junmai-shu is perfect during a meal. You can enjoy it with both warm and cold sake.

**醸造元：**藤井酒造株式会社
（1863年創業／広島県）
**種類：**日本酒 特別純米酒
**度数：**15%
**精米歩合：**65%

**Brewery** : Fujii Shuzo Co., Ltd.
（Since 1863／Hiroshima）
**Kind** : Special Junmai-shu
**Alcohol Content** : 15%
**Polishing Ratio** : 65%

**甘辛度**
Sweet-dryness

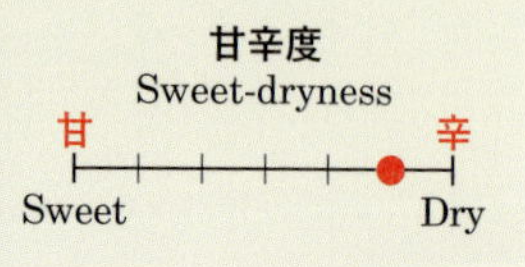

## 主な行き方 Major Directions

広島空港／バス＝(45分)⇒広島駅＝(30分)⇒宮島口駅／フェリー＝(10分)⇒宮島

Hiroshima Airport／Bus＝(45 min)⇒Hiroshima Station＝(30 min)⇒Miyajimaguchi Station／Ship ferry＝(10 min)⇒Miyajima

## 旅の宝 Treasure Hunt

### ▶①厳島神社

満潮時には海の上に浮かんでいるかのようにみえる厳島神社。今なお平清盛が造営したときと同様の姿のまま残っているといわれており、国宝に指定されています。海に囲まれたその姿は神秘的な印象であり、ライトアップされた夜間はさらにその歴史を感じさせます。

### ▶②大鳥居

宮島の象徴ともいえる朱色の大鳥居。満潮時には海に浮かんでいるような姿をみることができますが、干潮時には大鳥居まで歩いていくことも可能です。近くでみれば、鳥居の西側に月、東側には太陽の印を目にすることができ、陰陽道の影響とともに厳かな雰囲気を感じることができるでしょう。

### ▶①Itsukushimajinja Shrine

At high tide, the Itsukushimajinja Shrine reveals itself as if it is floating on the ocean. It is still remaining the form when it was first built by Kiyomori Taira, and is a designated National Treasure. The figure which is surrounded by the ocean is very mysterious and at nighttime when the shrine is lighten up, it makes us feel the history even more.

### ▶②The Great Torii

The red Great Torii is one of the symbols of Miyajima Island. At high tide, people can see the figure as if it is floating on the ocean, but at low tide people can walk to the Great Torii. When seen at a close distant, people can find a sign of the moon on the west side of the torii and a sign of the sun on the east side of it. People should be able to feel a solemn atmosphere along with the influence of Yin and Yang road.

**Brewery's recommended spot**
**蔵元おすすめスポット**

### 竹原の古い町並み

竹原市本町三丁目界隈
Hommachi, Takehara-shi

国の指定である伝統的建造物保存地域。

### Old Streets of Takehara

The country's registered important preservation districts for historical buildings.

## 紅葉谷公園

廿日市市宮島町紅葉谷
Momijidani, Miyajima-cho, Hatsukaichi-shi

名前の通り、モミジの木が何百本と植えられている、紅葉で有名な公園。11月中旬～下旬の紅葉の季節も赤く染まった見事な風景がみられるが、春から夏にかけての緑に染まる風景も清々しい。

## Momijidani Park

The park, from its name, has many maple leaves and is very famous for its autumn leaves. From the middle of November to the end, you can see the red leaves; however from spring to summer, you can see the green leaves which is very fresh.

## 弥山(みせん)

廿日市市宮島町
Miyajima-cho, Hatsukaichishi

宮島の中央にある、標高535mの山で、山頂は視界を遮るものがなく、360度広がる瀬戸内海の景色をみることができる。天気の良いときには、四国連山まで望めることも。2時間30分程の登山道で頂上を目指すこともできるが、観光には途中の獅子岩駅まではロープウエーで行くのがお勧め。

## Mt. Misen

Located in the middle of Miyajima, the mountain is 535 m high and from the pinacle there is nothing in the way, thus you can see the view of the setonaikai ocean clearly. When it is sunny, you can even see Shikoku. You can reach the pinacle by walking for about 2.5 hours, but you can take the rope way till Shishiiwa Station.

### 旅酒が見つかる場所 Shop

- **酒と器 久保田**
  廿日市市宮島町481(町家通り)
  0829-44-2228
- **広電 宮島ガーデン**
  広島市廿日市市宮島口1-11-7
  0829-56-0150
- **津田商店**
  広島市廿日市市宮島町95-1
  0829-44-0567
- **津田屋**
  広島市廿日市市宮島町843-3
  (表参道商店街)
  0829-44-0682

※最新の販売店情報はホームページでご確認ください。
Please check our website for the latest retail shop information.

# ISHIGAKIJIMA -Okinawa-

# 石垣島 （沖縄県）

05

沖縄離島の代表格である石垣島。誰もが憧れる南の島として人気を誇ります。サンゴ礁と海浜と、平野と森林と山、大自然の魅力をすべて網羅したこの地は、マリンスポーツからトレッキングといったスポーツとの相性も抜群。さらに独自の文化と沖縄料理を楽しむこともできる、まさに日本を代表する観光名所です。

It is famous for one of the tropical islands that everyone wants to go. The island contains all the attractions of great nature: the coral reef, the seaside, plains, forests, and mountains. It is suited for all kinds of marine sports and trekking. Also, the enjoyable peculiar culture and Okinawa food makes it one of the famous tourist resorts of Japan.

## 旅 TABI-SAKE 酒

### ISHIGAKIJIMA

5年貯蔵した古酒（クース）。古酒の香りとまろやかなのどごし。お米のみを原料としているため、お米の香り、味を感じられる泡盛になっています。

The old Okinawa's distilled spirits storage for 5 years. The scent of old sake and a mild feeling of drink. You can taste the flavor and scent of rice, as rice is the sole ingredient of this Awamori.

**醸造元**：株式会社玉那覇酒造所（1912年創業／沖縄県）
**種類**：琉球泡盛 5年古酒
**度数**：43度

**Brewery** ： Corporation Tamanaha Brewery（Since 1912／Okinawa）
**Kind** ：Ryukyu Awamori 5 years old sake
**Alcohol content** : 43%

## 主な行き方　Major Directions

羽田空港＝(190分)⇒南ぬ島 石垣空港
Haneda Airport＝(190 min)⇒Painushima Ishigaki Airport

成田国際空港＝(220分)⇒南ぬ島 石垣空港
Narita International Airport＝(220 min)⇒Painushima Ishigaki Airport

## 旅の宝　Treasure Hunt

### ▶①川平湾

日本百景にも選ばれた川平湾では、石垣島の魅力でもある海の美しさを存分に楽しめます。ボートで遊覧すれば、色とりどりの熱帯魚や珊瑚を見ることもできます。さらさらとした白い砂と、透明度が高くキラキラと輝く青い海のコントラストを楽しみましょう。

### ▶①Kabira Bay

At the Kabira bay, selected as the hundred views of Japan, people can surely enjoy the beauty of the ocean which is one of the attractions of Ishigaki Island. When people go sightseeing on a boat, they can see the various colors of tropical fish and the coral reef. Please enjoy the contrast of the white sand and the high transparency of brilliant blue ocean.

### ▶②唐人墓

船の座礁によって亡くなった中国人労働者のために建てられた唐人墓。中国人の霊を慰めるために建てられたこの墓は、竜や騎士などが装飾に施され、中国の文化を感じさせます。

### ▶②Toujin Grave

A grave built for the Chinese workers who have passed away by the sinking of the ship. The grave built to sympathize the souls of Chinese is decorated with knights and dragons and people can feel the Chinese culture.

### Brewery's recommended spot
### 蔵元おすすめスポット

## マングローブ群落

吹通川, 宮良川など
Fukidougawa River, Miyaragawa River etc. ...

干潮時にマングローブの群生が目の前で見られるスポット。

## The Mangrove Forest

The spot where you can see mangroves when the water is low.

## 旅酒が見つかる場所 Shop

- ㈲稲福酒販
  石垣市美崎町7-11
  0980-82-6363
- 石垣市特産品販売センター
  石垣市大川208 公設市場2F
  0980-88-8633
- 石垣島ショッピングプラザ
  石垣市平得370-3
  0980-82-5938
- とぅもーるショップ
  石垣市美崎町1
  石垣離島ターミナル内
  0980-88-0822
- うちなーみやげ館
  石垣市大川213-1
  0980-83-5118
- 島のみやげ館
  石垣市大川207-3
  0980-83-7111
- おみやげ本舗 なかそね家
  石垣市大川219
  0980-87-6296
- 玉那覇酒造所 店頭販売
  石垣市字石垣47
  0980-82-3165

※最新の販売店情報はホームページでご確認ください。
Please check our website for the latest retail shop information.

◆他にも素敵な場所があります。あなただけの旅の宝を探してみましょう!
**There are other wonderful places to visit. Let's take a trip to find a treasure of your own!**

06

## TAZAWAKO・KAKUNODATE -Akita-

# 田沢湖・角館 （秋田県）

日本百景にも選ばれている景勝地として有名な田沢湖。そしてみちのくの小京都として知られる角館。多くの観光客、宿泊客が訪れる人気スポットです。大地の恵みにあふれた食も魅力のひとつ。温泉と豊かな自然美とともに、田沢湖・角館を味わい尽くしてはいかがでしょうか。

Lakc Tazawa is a famous place where it is even selected as the Japan's hundred views. It is also known as little Kyoto. It is one of the places where many tourists visit. The rich land also grows great food. Please enjoy the beautiful nature and the hot springs in Lakc Tazawa.

## 旅 TABI-SAKE 酒

### TAZAWAKO・KAKUNODATE

東京農業大学の花酵母（撫子酵母）を使用し、天寿酒米研究会産契約栽培酒造好適米で醸し上げた、清んで華やかな香りと心やすらぐ味わいをお楽しみください。

Using the flower yeast (Nadeshiko yeast) from the Tokyo University of Agriculture, and a rice suited for making sake produced in a contract farm with a workshop class are brewed. Please enjoy the clean and glorious scent, and a flavor that relieves your heart.

**醸造元**：天寿酒造株式会社
（1830年創業／秋田県）
**種類**：日本酒 純米吟醸酒
**度数**：15度
**精米歩合**：60%

**Brewery**：TENJU SHUZO Co., Ltd.
（Since 1830／Akita）
**Kind**：Junmai Ginjo-shu
**Alcohol Content**：15%
**Polishing Ratio**：65%

**甘辛度**
Sweet-dryness

主な行き方 Major Directions

羽田空港＝(15分)⇒浜松町駅＝(5分)⇒東京駅＝(167分)⇒田沢湖駅＝(15分)⇒角館駅

Haneda Airport＝(15 min)⇒Hamamatsucho Station＝(5 min)⇒Tokyo Station＝(167 min)⇒Tazawako Station＝(15 min)⇒Kakunodate Station

### ▶①角館

みちのくの小京都と呼ばれる角館は、今なお城下町の面影が残っています。江戸時代にタイムスリップしたような気分を味わいたいのであれば、まずは武家屋敷通りと呼ばれるメインストリートに向かいましょう。角館は桜並木でも知られており、武家屋敷通りの両側に並ぶしだれ桜のなかを人力車で走れば、非日常感を味わうことができるでしょう。

### ▶①Kakunodate

Also known as little Kyoto, Kakunodate has many features of the castle town remaining. People can feel the atmosphere as if they have traveled back to Edo period. You should head to the main street called old samurai residences street first. Kakunodate is also known for the cherry blossoms; there are many trees at the side of the main street. You can ride the man powered taxi at the street and would be able to experience an unusual day.

### ▶②田沢湖

日本一の深さを誇る田沢湖。瑠璃色に光る湖面は、ブロンズのたつこ像とともにその美しさは変わることがありません。たつこ姫の伝説にあるように一年中凍らない湖として知られ、いつも来た人を楽しませています。

### ▶②Lake Tazawa

Lake Tazawa is known as the deepest lake in Japan. The surface of the like shining blue and the beauty of the bronze statue will never change. The legend of tatsuko princess is said that the lake will never freeze and would enjoy everyone at any time they visit.

**Brewery's recommended spot**
**蔵元おすすめスポット**

## 桧木内川の河川敷の桜並木

仙北市角館町岩瀬
Iwase, Kakunodate-machi, Semboku-shi

武家屋敷通りにほど近い桧木内川の川沿いには、全長2kmの桜が植えられている。夜はライトアップされて歩きながら桜を眺められるので、とても風情があり美しい。

## The row of cherry trees along the Hinokinai River

The cherry trees are planted along the Hinokinai River for about 2 km. The cherry trees are lighted up at night, thus you can enjoy seeing them at night; it is very refined and beautiful.

## 旅酒が見つかる場所 Shop

- **ワインと地酒 スガワラショップ**
  仙北市角館町雲然荒屋敷184-18
  0187-55-2242
- **田沢湖いち**
  仙北市田沢湖生保内字水尻59-40
  0187-43-2855
- **あきた芸術村 温泉ゆぽぽ 本館売店**
  仙北市田沢湖卒田字早稲田430
  0187-44-3905
- **あきた芸術村 温泉ゆぽぽ 温泉館売店**
  仙北市田沢湖卒田字早稲田430
  0187-44-3955
- **プラザホテル山麓荘**
  仙北市田沢湖生保内駒ヶ岳2-32
  0187-46-2131
- **斉藤商店**
  仙北市西木町西明寺字潟尻90
  0187-47-2381
- **花館商店**
  仙北市田沢湖生保内男坂130
  0187-43-0046
- **角館こだわり蔵**
  仙北市角館町横町42-1
  0187-52-1246
- **樹の下や よしなり**
  仙北市角館町表町下丁13-1
  0187-55-4133

※最新の販売店情報はホームページでご確認ください。
Please check our website for the latest retail shop information.

**◆他にも素敵な場所があります。あなただけの旅の宝を探してみましょう!**
**There are other wonderful places to visit. Let's take a trip to find a treasure of your own!**

# 磐梯 (福島県)

07

福島を代表する磐梯山。季節ごとに違った魅力を味わうことができるこの地域は、トレッキングからウォータースポーツ、スキー・スノーボードといったあらゆるアウトドアを楽しむことができます。豊かな自然の中、日常を忘れてリフレッシュしてみては？

Mt. Bandai represents Fukushima prefecture. At this area, people can enjoy different outdoor sports such as trekking, water sports, skiing, snowboarding in different seasons. Why don't you refresh yourself and forget your daily works in the abundant nature?

## 旅 TABI-SAKE 酒
### BANDAI

福島県産酒造好適米を原料として丁寧に醸した純米吟醸です。吟醸香をおさえた穏やかなやさしい香りと軽快でなめらかな味わい。きれいな旨味と酸味が際立ち、お燗でもおいしくお楽しみいただけます。

It is the Junmai Ginjo-shu, using the rice best for making sake from Fukushima prefecture. It has a light taste and a kind fragrance. It has a beautiful umami flavor and an acid taste; you can enjoy it by warming it as well.

**醸造元**：笹の川酒造株式会社
（1765年創業／福島県）
**種類**：日本酒 純米吟醸酒
**度数**：14度
**精米歩合**：60%

**Brewery** : SASANOKAWA SHUZO Co., Ltd.（Since 1765／Fukushima）
**Kind** : Junmai Ginjo-shu
**Alcohol Content** : 14%
**Polishing Ratio** : 60%

# 07 BANDAI

**Major Directions**

羽田空港＝（20分）⇒浜松町駅＝（5分）⇒東京駅＝（80分）⇒郡山駅＝（50分）⇒磐梯町駅

Haneda Airport＝(20 min)⇒Hamamatsucho Station＝(5 min)⇒Tokyo Station＝(80 min)⇒Koriyama Station＝(50 min)⇒Bandaicho Station

### ▶①磐梯山

日本の百名山にも選ばれている磐梯山は、表磐梯と裏磐梯とにわけられます。表磐梯では猪苗代湖畔や大蛇伝説のある不動滝などをみることができます。裏磐梯は火山の噴火によって生まれたことで、火口壁、五色沼、多くの湿地や沼の幻想的な雰囲気を楽しむことができます。

### ▶②喜多方ラーメン

喜多方ラーメンで全国的にその名を知られている喜多方。100軒以上のラーメン屋が軒を連ね、日本三大ラーメンと呼ばれています。コシのある太い縮れ麺をぜひご堪能ください。

### ▶①Mt. Bandai

Selected as the hundred mountains of Japan, Mt. Bandai is divided into front Bandai and back Bandai. At the front of Bandai, people can see the Inawashiro Lakeside and the Fudou Waterfall which is famous for the legend of a giant snake. On the other hand, at the back of Mt. Bandai, people can enjoy the wondrous atmosphere of the crater wall, Goshiki-numa Ponds, and many bogs and marshes which were all made by the eruption of the volcano.

### ▶②Kitakata Ramen

Kitakata is famous nationwide for the Kitakata Ramen. Over hundred ramen restaurants are opened and it is called the Japan's three biggest ramen. Please enjoy the thick chewy curly noodles.

## Brewery's recommended spot
## 蔵元おすすめスポット

### 三春滝桜

田村郡三春町滝桜久保296

296, Takisakurakubo, Miharu-machi, Tamura-gun

樹齢1000年超のベニシダレザクラの巨木で国の天然記念物である。四方に広げた枝から花が流れ落ちる滝のように咲くことから「滝桜」の名がついている。

### Miharu Takizakura

It is a huge red weeping cherry tree, thousand years old, and is the natural monument. It is named as "Takizakura" since the flower falls down like a waterfall from the branches spread to all sides.

## 旅酒が見つかる場所 Shop

- **四季彩一力**
  郡山市熱海町熱海4-161
  024-984-2115
- **清稜山倶楽部**
  郡山市熱海町熱海5-18
  024-984-2811
- **グランドサンピア猪苗代 リゾートホテル＆スキー場**
  耶麻郡猪苗代町綿場7126
  0242-65-2131

※最新の販売店情報はホームページでご確認ください。
Please check our website for the latest retail shop information.

**◆他にも素敵な場所があります。あなただけの旅の宝を探してみましょう!**
**There are other wonderful places to visit. Let's take a trip to find a treasure of your own!**

08

BEPPU -Oita-

# 別府 （大分県）

全国有数の温泉地として有名な大分県別府市。日本の約10分の1の源泉が集まる、世界でもトップクラスの温泉街です。源泉毎に異なる泉質を楽しみながら、日常の疲れを癒してはいかがでしょうか。

Beppu City, Oita Prefecture is famous for having many hot springs. About 1/10 of the source in Japan is concentrated here and it is one of the best Spa towns in the world. Why don't you relieve yourself with different kinds of spring qualities?

## 旅 TABI-SAKE 酒

BEPPU

雑味の元となる麦の筋を取り除くように磨き上げた“米粒麦”を原料に、旨みだけを抽出した味わい深い本格麦焼酎。旨みがギュッと凝縮されたクリアなおいしさです。

Using the “Kometsubumugi,” a type of ingredient that is made by removing the fibber of the wheat so that there is no unfavorable taste in shochu, the barely liquor is full of rich flavor. The umami flavor is condensed and is a clear taste.

**醸造元**：八鹿酒造株式会社
（1864年創業／大分県）
**種類**：本格麦焼酎
**度数**：25度

**Brewery**：Yatsushika Brewery Co., Ltd.（Since 1864／Oita）
**Kind**：Honkaku-Barley-shochu
**Alcohol Content**：25%

主な行き方 Major Directions

大分空港／バス＝(50分)⇒別府駅
Oita Airport／Bus＝(50 min)⇒Beppu Station

旅の宝 Treasure Hunt

### ▶①地獄めぐり

別府観光の定番中の定番、地獄巡り。8つの源泉を周遊する観光コースです。さまざまな景観を楽しめる地獄巡りは別府に行くなら外せないイベントです。

### ▶①Jigoku Meguri (Hell Tour)

Jigoku Meguri is the most common touring in Beppu. It is a tour which you visit around 8 different spring qualities. It is a must touring spot if you were to visit Beppu.

### ▶②湯けむり展望台

別府市街のいたるところにある温泉場を一望することができる展望台。湯けむりが立ち上る市街の様子は必見で、重要文化的景観にも指定されています。

### ▶②Yukemuri Sightseeing Tower

It is a Sightseeing Tower where you can see all the hot springs in Beppu City. The steam rising through the city is a must see and it is also registered as Japan's cultural landscape.

## Brewery's recommended spot
## 蔵元おすすめスポット

### 九重"夢"大吊橋

玖珠郡九重町田野1208

1208, Tano, Kokonoe-machi, Kusu-gun

高さ173m、長さ390mで、高さは日本一。スリルはあるが、絶景をみることができる。

### Kokonoe "Yume" Otsurihashi (suspension bridge)

The height is 173 m and the length is 390 m, it is the highest in Japan. There is thrill but you can enjoy the amazing view.

## 旅酒が見つかる場所 Shop

- **ホテル白菊 売店**
  別府市上田の湯町16-36
  0977-21-2111
- **別府交通センター**
  別府市新港町6-46
  0977-24-1811
- **JR九州 別府銘品蔵**
  別府市駅前町12-13
  0977-23-3653
- **大八**
  別府市松原町1-1
  0977-24-1414
- **別府ロープウェイ 九州焼酎館**
  別府市南立石字寒原10-7
  0977-22-2279
- **トキハ百貨店（別府店）**
  別府市北浜2-9-1
  0977-23-1111（代表）
- **高崎山おさる館**
  大分市神崎3078-40
  097-537-9294
- **さんふらわあマリンサービス 別府売店**
  別府市汐見町9-1
  別府観光港ターミナル1階
  0977-22-2898
- **もと湯の宿 黒田や**
  別府市鉄輪御幸3組
  0977-66-9656
- **龍巻地獄売店**
  別府市野田782 龍巻地獄
  0977-66-1854
- **大分銘品蔵**
  大分市要町1-40 JR大分駅
  豊後にわさき市場内
  097-513-7061

※最新の販売店情報はホームページでご確認ください。
Please check our website for the latest retail shop information.

◆**他にも素敵な場所があります。あなただけの旅の宝を探してみましょう!**
**There are other wonderful places to visit. Let's take a trip to find a treasure of your own!**

# BIWAKO -Shiga-

# 琵琶湖（滋賀県）

09

日本最大の湖で、滋賀県の面積のおよそ1/6を占める琵琶湖。クルージングやボート遊びなど、自然のなかでたくさんのレジャーが楽しめる湖です。
琵琶湖周辺には、彦根城など、貴重な歴史遺産も数多く、江戸時代からのお寺や神社を巡る、歴史ロマンを巡る旅も可能なエリアです。

As the largest lake in Japan, Lake Biwa accounts for about one-sixth of the area of Shiga prefecture. You can enjoy various leisure activities in nature such as cruising and boating there. Around Lake Biwa, there are many precious historical heritages including Hikone Castle. This area also has temples and shrines from the Edo period, where you can stroll while feeling a yearning for history.

## 旅 TABI-SAKE 酒
### BIWAKO

しっとりとしたやわらかな酸味が、うまみを引き出しました。後口のキレ味を意識し、上品にまとめました。

The light acid taste of the Special Junmai-shu bring the umami flavor out. It is made to have an elegant taste feeling the refreshing after-taste.

**醸造元**：喜多酒造株式会社
（1820年創業／滋賀県）
**種類**：日本酒 特別純米酒
**度数**：16度
**精米歩合**：60%

**Brewery** : Kita Shuzo Co., Ltd.
(Since 1820／Shiga)
**Kind** : Special Junmai-shu
**Alcohol Content** : 16%
**Polishing Ratio** : 60%

## 主な行き方 Major Directions

羽田空港＝（100分）⇒品川駅＝（20分）⇒米原駅＝（125分）⇒彦根駅＝（40分）⇒大津駅

Haneda Airport＝(100 min)⇒Shinagawa Station＝(20 min)⇒Maibara Station＝(125 min)⇒Hikone Station＝(40 min)⇒Otsu Station

### ▶①琵琶湖

日本最大・最古の湖。四季折々の風景を堪能することができるほか、夏には湖上クルーズ、ウォータースポーツを楽しむ大勢の人の姿が見られます。周長240kmにもなり、周辺にもさまざまな観光地があるので、ドライブもおススメです。

### ▶①Lake Biwa

The oldest and largest lake in Japan. Besides enjoying the scenery of each season, in summer you can cruise or do water sports. The circumference is about 240 km and there are many tour spots around the lake. Driving is also recommended.

### ▶②彦根城

国宝として指定されている彦根城天守閣は、現在でもほぼ当時の状態のままで残されています。建造物の美しさはもちろんのこと、春には1000本以上の桜が咲き誇り、彦根城の美しさを際立てます。

### ▶②Hikone Castle

Registered as the national treasure, the Castle Tower of Hikone Castle still remains its features from the past. The building, of course, is beautiful but in spring, over 1000 cherry blossoms bloom and the beauty of Hikone Castle stands out even more.

## Brewery's recommended spot 蔵元おすすめスポット

### 旧魚屋町

彦根市本町

Hommachi, Hikone-shi

彦根城の城下町として、今なお美しい町並みを残す旧魚屋町。多くの魚屋があったことからこの名がついたといわれ、今でも井戸を残している家や重厚な構えの市の指定文化財となっている家が残っている。

### Uoyamachi

As the Castletown of Hikone Castle, Uoyamachi still remains the beautiful townscape. Uoyamachi was named as it since there were many fish stores, and still there are houses with wells and many of them are the city's Designated Cultural Properties.

## 旅酒が見つかる場所 Shop

- **彦根キャッスル リゾート＆スパ 彦根みやげ本陣**
  彦根市佐和町1-8
  0749-21-3071
- **酒売処林屋**
  彦根市本町1-7-37
  0749-22-2737
- **彦根ビューホテル**
  彦根市松原町網代口1435-91
  0570-047-800

※最新の販売店情報はホームページでご確認ください。
Please check our website for the latest retail shop information.

**◆他にも素敵な場所があります。あなただけの旅の宝を探してみましょう!**
**There are other wonderful places to visit. Let's take a trip to find a treasure of your own!**

## MATSUYAMA・DOGO ONSEN -Ehime-

# 松山・道後温泉 （愛媛県）

日本最古の温泉のひとつとされている道後温泉。あの聖徳太子も訪れたというこの地は、日本人のみならず、海外からの観光客にも人気です。最近はアートの街としての評価も高く、多様な見所にあふれています。歴史的な温泉とあわせて、新しい魅力を味わってみてはいかがでしょうか。

Dogo Onsen is one of the oldest Spa resorts in Japan. This place, where the Shoutoku Taishi even visited, is famous not only from Japanese people, but also from foreign visitors. Recently, it is popular for the town of arts and is filled with various highlights. Why don't you enjoy the historical hot spring and the new attractions of Dogo?

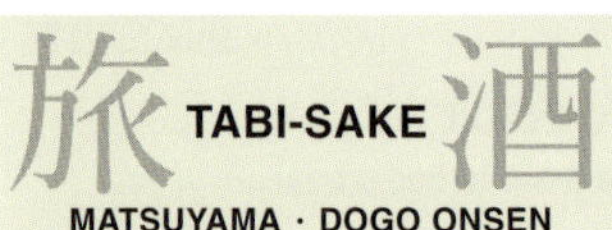

原酒ならではの濃厚な香りと、押しの強い旨味を持つ純米吟醸酒。
夏はロックで、冬にはお燗で幅広く楽しめるお酒です。

This is the Junmai Ginjo-shu with a rich aroma and full-bodied taste.
It can be enjoyed in a wide variety of ways, on the rocks in summer and hot in winter.

**醸造元：**梅錦山川株式会社（1872年創業／愛媛県）
**種類：**日本酒 純米吟醸酒
**度数：**16度以上17度未満
**精米歩合：**60%

**Brewery :** Umenishikiyamakawa Co., Ltd.（Since 1872／Ehime）
**Kind :** Junmai Ginjo-shu
**Alcohol Content :** 16%-17%
**Polishing Ratio :** 60%

## 主な行き方 Major Directions

羽田空港＝(65分)⇒伊丹空港＝(50分)⇒松山空港＝(15分)⇒松山駅＝(30分)⇒道後温泉駅

Haneda Airport＝(65 min)Itami Airport＝(50 min)⇒Matsuyama Airport＝(15 min)⇒Matsuyama Station＝(30 min)⇒Dogo Onsen Station

### ▶①放生園

放生園は道後温泉駅前にある広場です。市民、観光客からは足湯が憩いの場として親しまれています。また、坊ちゃんカラクリ時計も人気のスポットです。

### ▶②武家屋敷跡

湯築城跡である道後公園に復元されている武家屋敷は、当時の町並みを知ることのできるスポットです。時代を感じさせる町並みは、風情があり非日常感を味わうことができます。

### ▶①Hojoen Park

Hojoen Park is a public square located in Dogo Onsen station. Also, it is famous for Botchan, whichis the title of story written by Soseki Natsume, and the automaton clock.

### ▶②Old Samurai Residences

The old samurai Residences is restored at the Dogo Park, the remains of Yuchiku Castle, and people can understand the townscape of those days. The streets let you feel that history and extraordinary feelings.

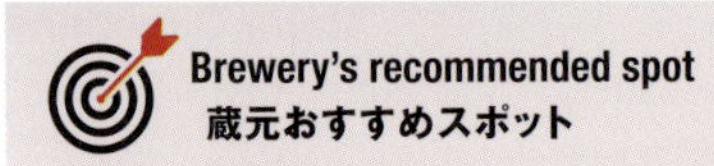


## 道後温泉本館

松山市道後湯之町5-6

5-6, Dogoyuno-machi, Matsuyama-shi

道後温泉のシンボル。現在の建物は、1894年に改築された木造3階建て。

## Dogo Onsen Honkan

It is the symbol of Dogo Onsen. The building right now was rebuilt in 1894 and is a 3 floor wooden building.

## 旅酒が見つかる場所 Shop

**●絣屋本店**
松山市道後湯之町20-19
089-941-3338

※最新の販売店情報はホームページでご確認ください。
Please check our website for the latest retail shop information.

**◆他にも素敵な場所があります。あなただけの旅の宝を探してみましょう!**
**There are other wonderful places to visit. Let's take a trip to find a treasure of your own!**

# YOKOHAMA -Kanagawa-

# 横浜 （神奈川県）

横浜は神奈川県の県庁所在地で東京の南に位置し、日本を代表する海の玄関口のひとつ。魅力的な商業施設と海と緑に囲まれた歴史名所を合わせ持つ日本有数の観光地です。港町ならではの異国情緒あふれる美しい景色や、中華街に代表されるようなさまざまな文化に触れることができ、多様な魅力を持つ町です。東京駅から30分前後で来られる好アクセスも魅力的です。

Yokohama is the capital of Kanagawa prefecture located south of Tokyo. Yokohama is one of the ports that represent Japan. It is famous for having many shopping malls, the ocean and green, and historical monuments. It is a city with a variety of attractions, with its exotic and beautiful scenery unique to a port city, as well as various cultures including Chinatown. Its convenient access with about 30 minutes from Tokyo Station is also tempting.

## 旅 TABI-SAKE 酒

### YOKOHAMA

神奈川の屋根、丹沢の伏流水の清らかな味とお米の芳醇な味のコラボレーション。冷やからぬる燗まで、温度帯で味わいの変化も楽しめます。

The collaboration of the clear flavor of the underground water of Tanzawa, called as the “roof” of Kanagawa, and the taste of rich rice. You can enjoy the Junmai-shu by both cold and warm; you can enjoy it with different temperatures, changing the taste slightly.

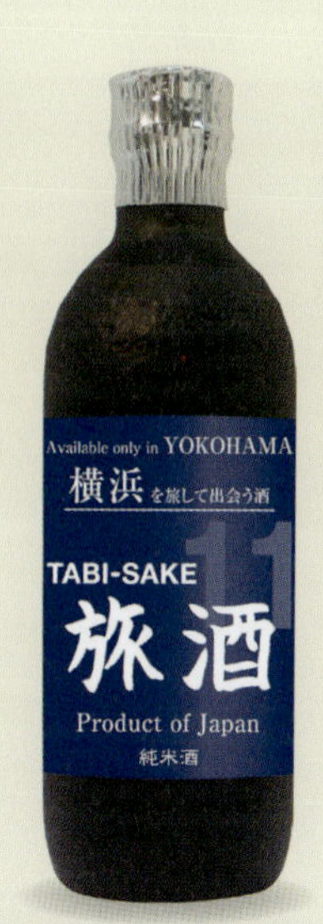

**醸造元：**有限会社金井酒造店
（1868年創業／神奈川県）
**種類：**日本酒　純米酒
**度数：**15%
**精米歩合：**60%

**Brewery** : Kanei Brewing Limited Company（Since 1868／Kanagawa）
**Kind** : Junmai-shu
**Alcohol Content** : 15%
**Polishing Ratio** : 60%

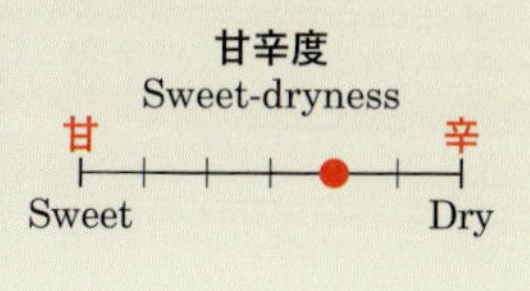

## 主な行き方 Major Directions

羽田空港＝(65分)⇒京急蒲田駅＝(15分)⇒横浜駅

Haneda Airport＝(65 min)⇒Keikyu Kamata Station＝(15 min)⇒Yokohama Station

## 旅の宝 Treasure Hunt

### ▶①横浜中華街

500軒以上のお店が集まる世界最大級の中華街、横浜中華街。年間2000万人以上の観光客が集まる、日本を代表する観光地のひとつにも挙げられています。

### ▶①Yokohama Chinatown

One of the largest Chinatown in the world, where almost 500 restaurants are gathered. Annually over 20 million tourists visit here.

### ▶②横浜赤レンガ倉庫

近代日本の発展を支えた港町横浜を代表する建造物、赤レンガ倉庫。明治から大正にかけて建造され、今では文化イベントなどが日々開催される人気観光スポットです。

### ▶②Yokohama Red Brick Warehouse

The red brick store is a famous building that represents Yokohama. It was built during Meiji to Taisho period, and is now a famous spot where cultural events are held.

**Brewery's recommended spot**
**蔵元おすすめスポット**

## 三渓園

横浜市中区本牧三之谷58-1
58-1, Sannoya, Hommoku, Naka-ku, Yokohama-shi

1906年に造園された17.5ヘクタールの敷地を持つ庭園。園内には、17棟の日本建築がある。
鑑桜の夕べなど、季節に応じた催しものを開催している。

## Sankeien Garden

It is a garden made in 1906 which has about a site of 17.5 hectare. Inside the garden, there are 17 Japanese buildings. There are events held for each season such as the Evening of the model cherry tree.

元町商店街は、江戸時代の横浜開港時に外国人向けの商店街として発展した街。いまなお、高級感のある洋服店や宝石用品店や老舗の洋菓子店・パン屋が軒を連ねている。買い物を楽しむのにおすすめのスポット。

## 元町商店街

横浜市中区元町
Motomachi, Naka-ku, Yokohama-shi

## Motomachi Shopping District

It was developed as a shopping area for foreign people when Yokohama Port opened in the Edo period. Even today, high-class clothing stores, jewelry stores, and long-established pastry shops and bakeries stand side by side along the streets. This is a recommended spot for shopping.

**旅酒が見つかる場所** **Shop**

- **東急ハンズ 横浜店**
  横浜市西区南幸1-3-1
  横浜モアーズ5～7階
  045-320-0109
- **Dining & Cafebar Living**
  横浜市西区高島2-11-11
  ポートビル3階
  045-628-9964
- **大ど根性ホルモン**
  横浜市西区北幸2-13-7
  第7NY ビル地下1階
  045-320-3077
- **和んず**
  横浜市中区桜木町
  ぴおシティ地下2階
  045-264-9688

※最新の販売店情報はホームページでご確認ください。
Please check our website for the latest retail shop information.

12

# CHICHIBU・NAGATORO -Saitama-

# 秩父・長瀞 (埼玉県)

都市部から電車で2時間、秩父・長瀞は自然と緑で満たされています。多くの観光客が一年中訪れる地域です。四季折々の自然とともに、旅酒をお楽しみください。

Although only 2 hours on train from the downtown, Chichibu and Nagatoro are filled with nature and green. Many tourists visit there all year long. It is a place where you can not only enjoy rafting but also the food such as natural water and shaved ice, and hot springs.

## 旅 TABI-SAKE 酒

### CHICHIBU・NAGATORO

吟醸香と純米の旨みを合わせ持つお酒です。50%まで磨かれた酒米で、上品な米の旨みが広がる味わい。常温か冷やしてお召し上がりください。

A sake both having the umami flavor of Ginjoko and pure rice. The sake rice is polished up to 50% and the taste of rice is refined. Enjoy the Junmai Daiginjo-shu at room temperature or cooling it.

**醸造元：**有限会社藤橋藤三郎商店（1848年創業／埼玉県）
**種類：**日本酒　純米大吟醸酒
**度数：**15度
**精米歩合：**50%

**Brewery :** Fujihashi Tosaburo Shoten Co., Ltd.（Since 1848／Saitama）
**Kind :** Junmai Daiginjo-shu
**Alcohol Content :** 15%
**Polishing Ratio :** 50%

## 主な行き方 Major Directions

羽田空港＝(20分)⇒浜松町駅＝(5分)⇒東京駅＝(40分)⇒熊谷駅＝(75分)⇒長瀞駅＝(20分)⇒秩父駅

Haneda Airport＝(20 min)⇒Hamamatsucho Station＝(5 min)⇒Tokyo Station＝(40 min)⇒Kumagaya Station＝(75 min)⇒Nagatoro Station＝(20 min)⇒Chichibu Station

## 旅の宝 Treasure Hunt

### ①ラフティング

長瀞でのレジャーで最も有名な物のひとつがラフティング。12月から3月中旬まで開催されています。時には激しく、時にはゆっくりと進む川の流れはエキサイティング。春は桜、夏は緑、秋は紅葉、冬はこたつ舟と、四季折々の景観を楽しむことができます。

### ②長瀞岩畳

天然記念物として登録されている長瀞岩畳。隆起した結晶片岩が文字どおり岩畳となって広がっており、対岸には秩父赤壁と呼ばれる絶壁や明神の滝をみることができます。

### ①River-rafting

River-rafting the Arakawa River is one of the most famous things to do in Nagatoro. Listening to the captains stories are very interesting features of the rafting. It is held from the middle of March to early December. The rafting goes sometimes very fast but sometimes slow; the speed fluctuation of the river excites people. For each season the rafting changes its feature for more excitement; in spring the cherry blossoms, in summer the green leaves, in fall the red leaves, and in winter the kotatsu boats.

### ②Nagatoro Iwadatami Rocks

It is registered as the natural monument. Like a tatami, the crystals are risen and lined along which makes you feel the greatness of nature. At the shore, you can see the Chichibu red wall and the waterfall. There are many other geographical spots to tour.

## Brewery's recommended spot
## 蔵元おすすめスポット

### 秩父夜祭

秩父市番場町1-1
1-1, Banba-machi, Chichibu-shi

秩父神社の例祭で、ユネスコ無形文化遺産に登録されている。山車の曳き回し、冬の花火大会で知られている。毎年12/1～12/6に行われる。

### Chichibu Yomatsuri Night Festival

It is an annual festival of Chichibujinja Shrine and is registered as the UNESCO's Intangible cultural heritage. Pulling around the float and the firework festival in winter are famous. It is held in December 1 to 6 ever year.

## 旅酒が見つかる場所 Shop

- **一本屋**
  秩父郡長瀞町長瀞459
  0494-66-0158
- **ちちてつ長瀞駅売店**
  秩父郡長瀞町長瀞528-2
  048-525-2283
- **SLパレオエクスプレス車内販売**
  株式会社秩鉄商事
  048-525-2283
- **西武秩父駅前温泉 祭の湯**
  秩父市野坂町1-16-15
  0494-22-7111

※最新の販売店情報はホームページでご確認ください。
Please check our website for the latest retail shop information.

◆**他にも素敵な場所があります。あなただけの旅の宝を探してみましょう!**
**There are other wonderful places to visit. Let's take a trip to find a treasure of your own!**

## GINZAN ONSEN -Yamagata-

# 13 銀山温泉（山形県）

大正ロマンが色濃く残る山あいの温泉地、銀山温泉。レトロな木造多層建築の旅館と「ガス灯」が織りなす幻想的な光景が魅力で、多数の観光客を虜にしている人気観光地です。夜景の人気も高く、ぜひ宿泊してその魅力を味わっていただければと思います。

Ginzan Onsen is a spa resort which the Taisho Romance still strongly remains. The wondrous view created by the nostalgic wooden multi storied buildings and the gaslight attracts many visitors. The night view is popular as well; please stay and enjoy the attractions.

## 旅 TABI-SAKE 酒

### GINZAN ONSEN

すっきりとしたやや辛口の純米大吟醸、口に含めば穏やかな香りが広がり、心地よい余韻を感じます。脂ののった秋刀魚やブリの塩焼きなど、魚料理と相性抜群です。

The Junmai Daiginjo-shu is a refreshing and dry taste; when you have it in your mouth, the fragrance spreads and makes you feel the comfortable. The sake goes well with fish dishes such as grilled pacific saury and yellowtail.

**醸造元**：和田酒造合資会社
（1797年創業／山形県）
**種類**：日本酒 純米大吟醸酒
**度数**：16度
**精米歩合**：50%

**Brewery** : Wada brewing jointstock company（Since 1797／Yamagata）
**Kind** : Junmai Daiginjo-shu
**Alcohol Content** : 16%
**Polishing Ratio** : 50%

甘辛度
Sweet-dryness

甘 Sweet ——————●—— 辛 Dry

## 主な行き方 Major Directions

羽田空港＝(20分)⇒浜松町駅＝(5分)⇒東京駅＝(195分)
⇒大石田駅／バス＝(35分)⇒銀山温泉

Haneda Airport＝(20 min)⇒Hamamatsucho Station＝(5 min)⇒Tokyo Station＝(195 min)Oishida Station/Bus(35 min)⇒Ginzan Onsen

### ▶①徳良湖

人工の湖である徳良湖は、キャンプ場やゴルフ場などのアウトドア施設が充実した市民の憩いの場となっています。湖の周辺に咲き誇る桜は約200本もあり、観光客を楽しませてくれます。

### ▶①Lake Tokurako

Lake is an artificial lake where camping areas, golf courses, and many other outdoor facilities enrich and citizens come to relax. There are about 200 cherry blossoms trees around the lake and they amuse the visitors.

### ▶②延沢銀坑洞

銀山温泉の地名の由来となった銀山跡です。江戸時代に幕府直営の鉱山として繁栄した名残を今に伝えています。 坑道内を探検することができるため、多くの観光客に人気のスポットです。

### ▶②Nobesawa Silver Mine

The trace of silver mine which is also the origin of Ginzan Onsen. In the Edo period, it was popular as a mine for the shogunate. Visitors can enjoy exploring the mining gallery and is popular among them.

**Brewery's recommended spot**
**蔵元おすすめスポット**

## ひな祭り(4月2、3日)

紅花資料館
西村山郡河北町谷地戊1143
Benibana (Satflower) Museum
1143, Yachibo, Kahoku-cho, Nishimurayama-gun

古くから紅花の産地として栄え、2018年に日本遺産登録された地域の中にある。紅花を介して京都の京雛が多数伝来しており、今でも旧暦でひな祭りを行っている。ひな市通りには出店が並び、旧家に飾られたお雛様を見学できる。紅花資料館では紅花の歴史や紅花商人の生活を垣間みることができる。

## Hina-matsuri (Doll Festival) (April 2,3)

It is located in the area which has prospered as a production area of safflower since ancient times. This region was designated as a Japanese heritage site in 2018. Many Kyobina dolls from Kyoto have been brought to the region through the safflower trade, and even now Hina-matsuri is held according to the lunar calendar. You can enjoy shopping at stalls along the Hina-ichi street and see the dolls displayed in old houses. At the Safflower Museum, you can get a glimpse of the history of safflower and the life of safflower merchants.

## 谷地どんが祭り(9月敬老の日の連休)

谷地八幡宮
西村山郡河北町谷地224
Yachi Hachimangu Shrine
224, Yachi, Kahoku-cho, Nishimurayama-gun

京都の祇園祭りが由来といわれる谷地八幡宮の例祭。豪華な囃子屋台が3日間町を練り歩く。1000年以上の歴史を持つ谷地八幡宮では、一子相伝といわれる谷地舞楽が奉納される。

問合せ先：一般社団法人河北町観光協会
TEL：0237-72-3787（代 表）／FAX：0237-73-3500
e-mail：info@benibananosato.jp
9:00-17:00（毎月第2木曜日、年末年始のぞく）

## Yachi Donga Festival

The Gion Festival of Kyoto is derived from the annual festival of Yachihachimangu Shrine. The glorius music band walks around the town for three days. The Yachibugaku, a mystery of art, is dedicated to the shrine, which has a history of over 1000 years.

## 旅酒が見つかる場所 Shop

- **八木橋商店**
  尾花沢市銀山新畑448
  0237-28-2035
- **さくらんぼ（山形空港ビル内）**
  東根市大字羽入字柏原新林3008
  0237-47-2111
- **銀山温泉「大正ろまん館」**
  尾花沢市大字上柳渡戸字十分一364-3
  0237-53-6727
- **大石田駅舎店**
  北村山郡大石田町駅前通り1-2
  0237-36-1515

※最新の販売店情報はホームページでご確認ください。
Please check our website for the latest retail shop information.

14

## ASAKUSA -Tokyo-

# 浅草 （東京都）

東京を代表する下町、浅草。都心からもほど近いこの地域に古き良き情緒あふれる町並みがあることは、日本を訪れる外国人観光客にとってもうれしい見所です。雷門や浅草寺は日本人、外国人ともに多くの観光客に人気なスポット。そんな浅草の魅力を味わい尽くしましょう。

Asakusa is the traditional shopping, entertainment, and residential districts that represent Tokyo. Close from the downtown, this area has good old atmosphere favored by Japanese people and foreign visitors. The Kaminarimon Gate and Sensoji Temple are famous tourist spots for visitors. Please enjoy Asakusa.

### 旅 TABI-SAKE 酒

ASAKUSA

米の旨みを引き出した ふくよかでコクのある味わい。やさしい香りを持ち、冷やから燗まで楽しめる純米酒。

Rich and well-rounded taste that brings out the full flavor of rice, the Special Junmai-shu with its gentle scent can be enjoyed either hot or cold.

**醸造元**：小澤酒造株式会社
（1702年創業／東京都）
**種類**：日本酒 特別純米酒
**度数**：15度
**精米歩合**：60%

**Brewery** : Ozawa Shuzo Co., Ltd.
（Since 1702／Tokyo）
**Kind** : Special Junmai-shu
**Alcohol Content** : 15%
**Polishing Ratio** : 60%

甘辛度
Sweet-dryness

## 主な行き方 Major Directions

羽田空港＝（40分）⇒浅草

Haneda Airport＝（40 min）⇒Asakusa

## 旅の宝 Treasure Hunt

### ▶①浅草寺

浅草の観光の中心地であるのが浅草寺です。浅草寺は東京最古の寺であり、いつも多くの人でにぎわっています。境内では五重塔や大仏などもみることができ、寺といえども見所十分です。

### ▶②花やしき

都心のなかに古くからある遊園地、それが花やしきです。花やしきは何といっても、日本最古のジェットコースターで知られています。古さと伝統を感じさせながらも、アトラクションとして気分を盛り上げてくれる乗り物がいっぱいです。

### ▶①Sensoji Temple

The main tour of Asakusa is the Sensoji Temple. Sensoji Temple is the oldest temple in Japan and is always filled with people. In the precincts, even though as a temple, you can see the five-storied pagoda, and a great image of Buddha.

### ▶②Hanayashiki

Hanayashiki is an amusement park that exists in downtown from long ago. Hanayashiki is famous for its oldest roller coaster in Japan. Even makin you feel the oldness and tradition, there are many rides to make you more excited.

Brewery's recommended spot
蔵元おすすめスポット

## 浅草三社祭(浅草神社例大祭)

台東区浅草2-3-1
2-3-1, Asakusa, Taito-ku

毎年5月に行われる浅草神社の例大祭。最大の特徴は、浅草中を練り歩く約100基の神輿。

## Asakusa Sanja Festival (Annual Festival of Asakusa-jinja Shrine)

It is an annual festival held in Asakusajinja Shrine every May. The greatest feature is the 100 Mikoshies walking around Asakusa.

## 東京スカイツリー

墨田区押上1-1-2
1-1-2, Oshiage, Sumida-ku

高さ634mで世界一高い電波塔。「空中散歩」をしている気分を楽しめる展望回廊はもちろん、日によって変わるエレベーターホールのライティングも楽しめる。

## Tokyo Sky Tree

It is the tallest broadcast tower in the world, having the height of 634 m. The observation floor makes you feel like walking the sky, and the lighting of the elevator changes every day so that people can enjoy it every day.

### 旅酒が見つかる場所 Shop

- **三平ストア 浅草店**
  台東区西浅草2-14-16
  03-5826-7203
- **四方酒店**
  台東区浅草1-19-10
  03-3841-6615
- **北野エース 松屋浅草店**
  台東区花川戸1-4-1
  松屋浅草地階
  03-3842-1370
- **あづまばし 四方酒店**
  台東区雷門2-1-14
  03-3844-0679

※最新の販売店情報はホームページでご確認ください。
Please check our website for the latest retail shop information.

（熊本県）

15

有名な活火山、阿蘇山をはじめとする雄大な自然と美しい湧水、神話にまつわる名所も多い阿蘇。巨大なカルデラも形成されており、その圧倒的なスケールは日本有数の観光スポット。九州旅行の際には訪れておきたい観光地です。

Aso is famous for its active volcano, Mt. Aso, as well as its magnificent nature, beautiful spring water and mythical places. Mt. Aso has a huge caldera, the overwhelming scale of which is one of the major tourist spots in Japan. It is a sightseeing place that you must visit when traveling in Kyushu.

## 旅 TABI-SAKE 酒

**ASO**

熊本県産「山田錦」を55%磨いて醸した純米吟醸酒。日本酒度マイナス20を目指した甘口でありながらもキレがあるタイプのお酒です。

It is the Junmai Ginjo-shu made from 55% of "Yamada Nishiki" from Kumamoto. Aiming for -20 degrees of sake, it has a sweet taste but at the same time has sharp tastes.

**醸造元：**河津酒造株式会社
（1932年創業／熊本県）
**種類：**日本酒　純米吟醸酒
**度数：**16度
**精米歩合：**55%

**Brewery：**KAWAZU SAKE BREWING COMPANY（Since 1932／Kumamoto）
**Kind：**Junmai Ginjo-shu
**Alcohol Content：**16%
**Polishing Ratio：**55%

**甘辛度**
Sweet-dryness

## 主な行き方 Major Directions

羽田空港＝(110分)⇒熊本空港／バス＝(70分)⇒阿蘇駅

Haneda Airport＝(110 min)⇒Kumamoto Airport/Bus＝(70 min)⇒ASO Station

### ▶①草千里ヶ浜

阿蘇観光に訪れたら必ず訪れたいスポット。阿蘇五岳の1つ、烏帽子岳の北麓に広がる火口跡にある大草原と、雨水が溜まってできたといわれる池のコントラストが美しい場所です。放牧された牛や馬と煙を上げる阿蘇山を望める絶景スポットで、草原を一周できる引き馬乗りでゆっくり散策するのも人気です。

### ▶②黒川温泉

九州だけでなく全国的にも有名な、緑ゆたかな山々に囲まれた黒川温泉郷。「黒川温泉一旅館」のコンセプトを持ち、黒川温泉郷全体が1つの旅館としてお客様をおもてなしし、訪れて良かった場所として、多くのお客様から高く評価されています。

### ▶①Kusasenrigahama

This is a sightseeing spot you definitely want to visit when you travel to Aso. It is a place with a beautiful contrast of the prairie in the crater site located to the north of the summit of Mt. Eboshi, one of the five peaks of Aso, and the pond which is said to have accumulated rainwater. It is a superb scenic spot where you can see grazing cows and horses with a flow of rising volcanic smoke from Mt. Aso behind them. It is also popular to go around the grassland by riding a horse at a leisurely pace.

### ▶②Kurokawa Onsen

Kurokawa Onsen surrounded by mountains with lush greenery is famous not only in Kyushu but also nationwide. With the concept of “Kurokawa Onsen Ichi Ryokan,” which literally means “Kurokawa hot spring town is one Japanese inn,” the whole town serves guests as if it were a single inn welcoming visitors. This onsen is highly appreciated by many tourists as a good place to visit.

**Brewery's recommended spot**
**蔵元おすすめスポット**

### 鍋ケ滝

阿蘇郡小国町黒渕

Kurobuchi, Oguni-machi Aso-gun

滝幅約20mあり。カーテンのように幅広く落ちる滝の水が神秘的。

### Nabegataki Waterfall

The waterfall has a width of 20 m and looks like a curtain of water.

## 下城滝(大銀杏木)

阿蘇郡小国町下城
Shimojo, Oguni-machi Aso-gun

落差約40mを一気に流れ落ちる豪快な滝。滝の近くにある、樹齢1000年以上という大イチョウも有名。

## Shimonjo-no-taki Waterfall (Large Ginkgo Tree)

Has a height of 40 m and the water falls dynamically from the top. The 1000 years old ginkgo tree near the waterfall is famous as well.

## 北里柴三郎記念館

阿蘇郡小国町北里3199
3199, Kitazato, Oguni-cho, Aso-gun

新千円札のデザインに選ばれた北里柴三郎博士は、阿蘇郡小国町の出身。博士から小国町に寄贈された北里文庫（図書館）を改修し、その偉業をたたえているのが北里柴三郎記念館。

## Kitazato Shibasaburo Memorial Hall

Dr. Shibasaburo Kitazato, whom the new 1,000 yen note will feature, was born in Oguni-cho, Aso-gun. Dr. Kitazato donated the Kitazato Bunko (library) to Oguni-cho, and it has been renovated to the Kitazato Shibasaburo Memorial Hall to honor his achievements.

### 旅酒が見つかる場所 Shop

- **後藤酒店**
  阿蘇郡南小国町大字満願寺6991-1
  0967-44-0027
- **マルサン 児玉酒店**
  阿蘇郡小国町宮原1685
  0967-46-2003
- **黒川温泉 山あいの宿 山みず木**
  阿蘇郡南小国町万願寺6392-2
  0967-44-0336
- **黒川温泉 旅館 こうの湯**
  阿蘇郡南小国町満願寺6784
  0967-48-8700
- **源流の宿 帆山亭**
  阿蘇郡南小国町満願寺6346
  0967-44-0059
- **河津酒造 小売り部**
  阿蘇郡小国町宮原1734-2
  0967-46-2311
- **河津酒造 地酒屋**
  阿蘇郡小国町宮原1763-3
  0967-46-6151

※最新の販売店情報はホームページでご確認ください。
Please check our website for the latest retail shop information.

16

# OIRASE & TOWADA LAKESIDE -Aomori-

# 奥入瀬十和田湖畔 (青森県)

新緑や紅葉、四季それぞれで違った顔をみせ、人気も高い奥入瀬・十和田湖畔。雄大な自然が織りなす美しい風景は、季節をかえて何度も訪れたくなる魅力を持っています。国指定の特別名勝および天然記念物として、多くの観光客が訪れています。

Oirase Towada Lakeside is famous for its fresh green, red leaves, and different features in each season. The beautiful view of the great nature is an attraction that people want to see in different seasons. Many tourists visit for the National Site of Scenic Beauty and the natural monument.

## 旅 TABI-SAKE 酒

### OIRASE & TOWADA LAKESIDE

桃川酵母と青森県産酵母の2つの酵母を使用して醸すことにより、豊かな吟醸香と飲みごたえのある味わいを両立。奥入瀬川水系の軟水の仕込水による口あたりの柔らかさとの全体のバランスの良いお酒です。

The Junmai Ginjo-shu is using yeast from Momokawa and Aomori, thus it has a Ginjo-shu fragrance and a thick taste. The sake is very balanced and the soft water of Oirase River makes it a soft texture when held in your mouth.

**醸造元**：桃川株式会社
（1889年創業／青森県）
**種類**：日本酒 純米吟醸酒
**度数**：15度以上16度未満
**精米歩合**：60%

**Brewery** : Momokawa Co., Ltd.
(Since 1889／Aomori)
**Kind** : Junmai Ginjo-shu
**Alcohol Content** : 15%-16%
**Polishing Ratio** : 60%

## 主な行き方 Major Directions

羽田空港＝(80分)⇒三沢空港＝(15分)⇒三沢駅＝(20分)⇒八戸駅＝(100分)⇒石ケ戸／徒歩＝(2分)⇒奥入瀬渓流＝(2分)⇒石ケ戸／徒歩＝(6分)⇒十和田湖

Haneda Airport＝(80 min)⇒Misawa Airport＝(15 min)⇒Misawa Station＝(20 min)⇒Hachinohe Station＝(100 min)⇒Ishigedo Station/Walking＝(2 min)⇒Oirase Stream＝(2 min)⇒Ishigedo Station/Walking＝(6 min)⇒Lake Towada

### ▶①奥入瀬渓流

十和田湖畔から焼山まで約14km続く日本を代表する景勝地。渓流沿いに数多くの滝が点在しており、「瀑布街道」とも呼ばれています。天然記念物として国の保護も受けており、多くの見所があるうえに探索のしやすさも人気です。

### ▶①Oirase Streem

Oirase Streem, a scenic spot that represents Japan, lasts 14 km from Towada Lakeside to Mt. Yakeyama. Many waterfalls are dotted along the stream and is called the Bakufukaido. As a natural monument it is receiving preservation from the country and is famous for the variety of features people can explore.

### ▶②十和田湖

青森と秋田にまたがる、その自然美が名高い十和田湖。春の桜、夏の新緑、秋の紅葉、冬の雪景色、四季折々の美しい風景が魅力です。展望台や遊覧船で、多彩な表情をみせる十和田湖を楽しむことができます。

### ▶②Lake Towada

Lake Towada, locating between Aomori Pretecture and Akita Pretecture, is famous for its natural beauty. The cherry blossoms in spring, fresh green in the summer, red leaves in fall, and winter views in winter are all the beautiful scenes people can enjoy. People can enjoy different features of the lake from observation platforms and sightseeing boats.

**Brewery's recommended spot**
**蔵元おすすめスポット**

## 十和田神社パワースポット「占場」

十和田市奥瀬十和田湖畔休屋486
486, Towadakohanyasumiya, Okuse, Towada-shi

以前は徒歩でも行くことができたが、現在は遊覧船かボートでしか行けない。十和田神社で「おより紙」をいただき、願いごとをしながら「占場」で湖に浮かべ、「おより紙」が水中に引き込まれるように沈めば、願いが叶うといわれている。

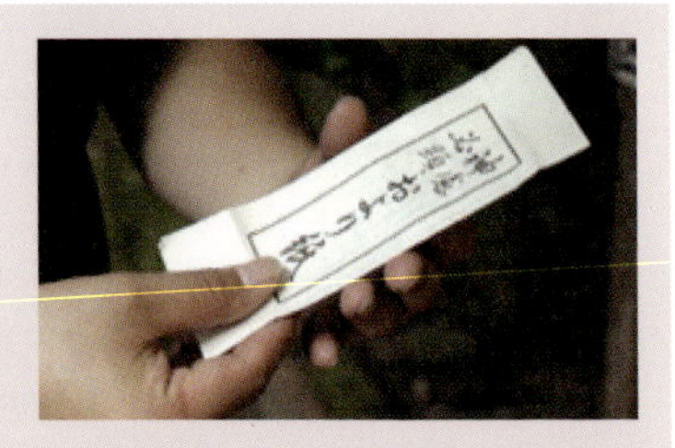

## Towadajinja Shrine's "Uranai-ba"

In the past, people were able to walk over there; however, people can only take a boat or a cruiser to go there. After receiving a Oyori-gami at the Towada-jinja Shrine, make a wish a at the Uranai-ba and float the paper on the water. It is said that if the Oyori-gami is pulled into the water, the wishes come true.

## 旅酒が見つかる場所 Shop

● **ホテル十和田荘**
十和田市奥瀬十和田湖畔休屋340
0176-75-2221

● **渓流の駅 おいらせ**
十和田市大字奥瀬字枋久保11-12
0176-74-1121

● **道の駅奥入瀬 四季彩館**
十和田市大字奥瀬字塘道39-1
0176-72-3201

※最新の販売店情報はホームページでご確認ください。
Please check our website for the latest retail shop information.

◆**他にも素敵な場所があります。あなただけの旅の宝を探してみましょう!**
**There are other wonderful places to visit. Let's take a trip to find a treasure of your own!**

## HAKATA -Fukuoka -

17

# 博多 (福岡県)

遥か昔から繁栄してきた九州の一大都市博多。大陸との関係も深く、古くから交易が盛んでもあり、日本の歴史に大きく影響してきた町。多くの魅力を持つこの町は、国内外から多くの観光客が訪れる地域にもなっています。

Hakata is one of the largest cities in Kyushu that has been thriving from the past. It is a city that has been affecting Japan's history since it has a strong relationship with the main island and is prosperous for trading. This city which has many attractions is popular among visitors from both inside and outside of Japan.

### 旅 TABI-SAKE 酒

HAKATA

宝満山の伏流水で仕込んだ純米吟醸。余韻がぼやけず安定したコシがある味。
切れ味の良い辛口で飲みあきしません。
冷やで切れが冴え、燗で豊かな味わいとなります。

Brewed with the underground water of Mt. Homan, this Junmai Ginjo-shu has a full-bodied flavor with a clear aftertaste. You never get tired of the dry and crisp taste. You can enjoy its sharp taste at room temperature, and an even richer taste by warming it.

醸造元：大賀酒造株式会社
（1673年創業／福岡県）
種類：日本酒 純米吟醸酒
度数：15度
精米歩合：55%

**Brewery** : Oga Shuzo Co., Ltd.
(Since 1673／Fukuoka)
**Kind** : Junmai Ginjo-shu
**Alcohol Content** : 15%
**Polishing Ratio** : 55%

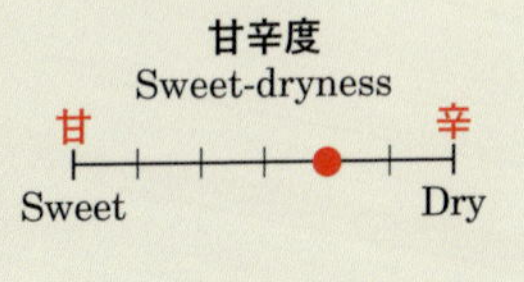

## 主な行き方 Major Directions

羽田空港＝（80分）⇒福岡空港＝（5分）⇒博多駅

Haneda Airport＝（80 min）⇒Fukuoka Airport＝（5 min）⇒Hakata Station

## 旅の宝 Treasure Hunt

### ▶①櫛田神社

地元では「お櫛田さん」の愛称で親しまれている櫛田神社。境内には大銀杏の御神木があり、博多のシンボルのひとつとなっています。博多祇園山笠や博多おくんちなどの例祭も有名です。

### ▶①Kushidajinja Shrine

The Kushidajinja Shrine is called Okushida-san by the local people. In the area, there is a big ginkgo tree, a sacred tree, which is one of Hakata's symbols. The Hakata Gion Yamakasa and the Hakata-Okunchi are also famous annual festivals in Hakata.

### ▶②博多祇園山笠

国の重要無形民俗文化財に指定されている祇園祭、博多祇園山笠。700年以上の伝統のある祭りで、全国から観光客が集まる博多の一大イベントです。

### ▶②Hakata Gion Yamakasa Festival

The Gion festival is designated as the country's Important Intangible Folk Cultural Property. It has a history of over 700 years and many tourists all over Japan visit this event.

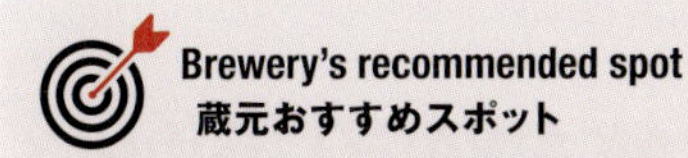

## 新元号「令和」ゆかりの地、太宰府市の「坂本八幡宮」

太宰府市坂本3-14-23

3-14-23, Sakamoto, Dazaifu-shi

坂本八幡宮は古代の歌人である大伴旅人の邸宅があったといわれる場所。新元号「令和」は大伴旅人の屋敷で開かれた梅の花をめでる宴で詠まれた万葉集の歌の序文からの引用。

## Sakamoto-hachimangu Shrine

There used to be a residence for Otomo no Tabito, a poet of ancient times, at this place, where the shrine is located today. The new era name “Reiwa” was taken from one poem in an ancient anthology, *Manyoshu*, and the poem is said to have been written when he hosted a plum-blossom party at his residence.

## 全国でも珍しい恋の神様、水田天満宮の末社「恋木神社」

筑後市水田62-1

62-1, Mizuta, Chikugo-shi

「良縁幸福の神様」として、親しまれている珍しい神社。境内には、ハート陶板が配置された恋参道がある。

## Koinokijinja Shrine

The Koinokijinja Shrine is a rare shrine which serves a god for love. In the area, there is a love street where the heart shaped ceramic tile is placed.

### 旅酒が見つかる場所 Shop

- **IWATAYA 岩田屋 本館**
  福岡市中央区天神2-5-35
  092-726-3787
- **おみやげ本舗博多**
  福岡市博多区博多駅中央街1-1
  092-475-0642

※最新の販売店情報はホームページでご確認ください。
Please check our website for the latest retail shop information.

18

## HIDATAKAYAMA SHIRAKAWAGO -Gifu-

# 飛騨高山白川郷 （岐阜県）

城下町としての面影を色濃く残す飛騨高山。広い敷地に手つかずの自然が残されており、集落全体が世界遺産となっている白川郷。ともに世界中から観光客が訪れる名所です。

Hidatakayama is a mountain located in the north part of Gifu prefecture. A lot of nature is still remaining in large areas, and Shirokawagou, which all the community is the world's prefecture, is a famous touring spot where people from all round the world visit.

## 旅 TABI-SAKE 酒

### HIDATAKAYAMA SHIRAKAWAGO

酒もろみをキレのある純米酒にささやかに添えることで、口のなかで米の味や香りが広がります。スッキリとした品のある余韻を残して、のどを過ぎる味わいに仕上がっています。清酒感覚で幅広い料理と合わせてお楽しみいただけます。

This Junmai-shu which has crisp taste was added to liquor mash. Therefore, the rice flavor spreads in your mouth, and after taste is refreshing and light. You can enjoy at ease, and with many kinds of dish.

**醸造元**：株式会社三輪酒造
（1837年創業／岐阜県）
**種類**：日本酒 純米酒
**度数**：15.5度
**精米歩合**：60%

**Brewery** : Miwa Shuzo Co., Ltd.
(Since 1837／Gifu)
**Kind** : Junmai-shu
**Alcohol Content** : 15%
**Polishing Ratio** : 60%

**甘辛度**
Sweet-dryness

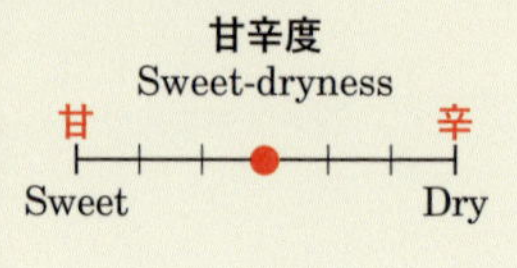

## 主な行き方 Major Directions

羽田空港＝(20分)⇒品川駅＝(100分)⇒名古屋駅＝(30分)⇒高山駅＝(90分)⇒猪谷駅＝(50分)⇒富山駅＝(20分)⇒高岡駅／バス＝(130分)⇒白川郷

Haneda Airport＝(20 min)⇒Shinagawa Station＝(100 min)⇒Nagoya Station＝(30 min)⇒Takayama Station＝(90 min)⇒Inotani Station＝(50 min)⇒Toyama Station＝(20 min)⇒Takaoka Station/Bus＝(130 min)⇒Shirakawa-go

## 旅の宝 Treasure Hunt

### ▶①高山祭り

16世紀後半から17世紀が起源とされる高山祭は春の「山王祭」と秋の「八幡祭」のふたつの総称です。祭行事ではにぎやかな伝統芸能も繰り広げられます。

### ▶①Takayama Festival

The Takayama Festival originated between the late 16th centuries and the 17th centuries. Its generic name came from the Sanno Festival held in spring and the Hachiman Festival held in Autumn. In the festival, lively traditional performing are held.

### ▶②白川郷

合掌造りの集落で知られ、独特の景観をなす白川郷。世界文化遺産にも登録され、毎年多くの観光客がこの地を訪れます。

### ▶②Shirakawa-go

It is known for a village with thatched roof houses and has a peculiar landscape. It is also registered as the world's cultural heritage and many tourists visit this area every year.

## Brewery's recommended spot
## 蔵元おすすめスポット

### 高山陣屋

高山市八軒町1-5

1-5, Hachiken-machi, Takayama-shi

江戸時代に飛騨が徳川幕府直轄地になったときに、江戸から来た役人が政治を行った役所。全国でも残っているのは高山だけ。

### Takayama Jinya

In the Edo period, when Hida became Tokugawa's area, it was a place where the governmental officers from Edo came to do politics. Takayama is the only place left in Japan.

## 旅酒が見つかる場所 Shop

- **道の駅 白川郷**
  大野郡白川村飯島411
  05769-6-1310
- **今藤商店**
  大野郡白川村荻町226
  05769-6-1041
- **飛騨特産工産物さくら井や**
  高山市花里町5-20
  0577-34-0931

※最新の販売店情報はホームページでご確認ください。
Please check our website for the latest retail shop information.

◆**他にも素敵な場所があります。あなただけの旅の宝を探してみましょう!**
**There are other wonderful places to visit. Let's take a trip to find a treasure of your own!**

# NAGASAKISHI・SHIMABARA -Nagasaki-

# 長崎市・島原 （長崎県）

19

コンパクトな都市の中に数多くの観光名所が存在する長崎市。江戸時代、鎖国のなかでも出島を通して欧州文化が伝わっていたことから生まれた独自の文化は、他の地域ではなかなか味わえません。豊かな自然に恵まれた島原半島のなかでも、島原市は地下水による湧水に恵まれ、「水の都」ともいわれています。江戸時代からの街並みがそのまま残る城下町でもあり、散策するスポットはたくさんあります。そんな長崎市・島原市の魅力を堪能していきましょう。

Nagasaki city, a compact city, has many tourist sites. During the edo period, even though Japan was a closed country, via Dejima, the european culture spread, and Nagasaki has a peculiar culture that you cannot experience in other areas. Out of all the Shimabara Peninsula which is full of nature, the Shimabara city is rich in underground water and is called as the city of water. The townscape is remaining from the Edo period and there are many spots to look around. Please enjoy the attractions of Nagasaki and Shimabara.

## 旅 TABI-SAKE 酒

### NAGASAKISHI ・ SHIMABARA

名水百選の雲仙島原の天然水を仕込水にし、フルーティーな香りとほど良いコクとふくらみのある味わいが特徴の、あと口のキレがいい純米吟醸酒です。

Using the famous water of Unzen Shimanbara, it has a fruity taste and a little of body, also it has a refreshing and light taste.

**醸造元**：合資会社山崎本店酒造場（1884年創業／長崎県）
**種類**：日本酒　純米吟醸酒
**度数**：15度
**精米歩合**：55%

**Brewery** : Joint-stock Company Yamazaki Head Office Sake Brewery（Since 1884／Nagasaki）
**Kind** : Junmai Ginjo-shu
**Alcohol Content** : 15%
**Polishing Ratio** : 55%

甘辛度
Sweet-dryness

## 主な行き方 Major Directions

羽田空港＝（120分）⇒長崎空港／バス＝（100分）⇒島原駅
Haneda Airport＝(120 min)⇒Nagasaki Airport／Bus＝(100 min)⇒Shimabara Station

## 旅の宝 Treasure Hunt

### ▶①教会群とキリスト教関連遺産

キリスト教の伝来から、日本国内で激動の歴史を経験した長崎。世界遺産候補に上げられているものも多く、象徴的な存在である大浦天主堂をはじめ、複数の教会を巡ることができます。

### ▶①Churches and inheritance related to the Christian culture

After the Introduction of Christian, Nagasaki experienced a tremendous change in the Japanese history. There are many building nominated as the World's inheritance; starting with the symbolic Oura-tenshudo, people can visit many varieties of churches.

### ▶②グラバー園

開国前の日本に訪れた実業家、トーマス・グラバー。彼が建てた南山手の住居は日本史好きには外せない観光名所です。

### ▶②Glover Garden

Before Japan opened its country, Thomas Glover, a businessman, visited Japan. The house he built in Minami Yamate is a famous place where people who love Japanese history should definitely visit.

## Brewery's recommended spot
## 蔵元おすすめスポット

### 湧水庭園 四明荘

島原市新町2-12
2-12, Shimmachi, Shimabara-shi

1日約1000トンの湧き水量を誇る池には色とりどりの鯉が泳ぐ。
建物は、明治後期の建築物で、国の登録有形文化財に登録されている。

### Yusui Garden Shimeiso

The water of the pond springs about 1000 tons per day and many kinds of carp are swimming. The building, built in the late Meiji period, is a country's Registered Tangible Cultural Property.

## 旅酒が見つかる場所 Shop

- **すみや**
  長崎市尾上町1-1
  アミュプラザ長崎1階
  095-808-3833
- **吉田酒店**
  長崎市恵美須町7-25
  095-821-1695
- **ショッピングセンター NAGASAKI**
  長崎市新地町9-1
  095-828-1877
- **稲佐山観光ホテル レストラン珍陀亭**
  長崎市曙町40-23
  095-861-4151
- **ホテルニュー長崎 長崎名店街 長崎お土産 すみや**
  長崎市大黒町14-5
  095-828-3733
- **長崎物産館**
  長崎市大黒町3-1
  交通産業ビル2階
  095-821-6580
- **島原城本丸売店**
  島原市城内1丁目1182
  0957-63-4874
- **雲仙湯守の宿 雲仙湯元ホテル**
  雲仙市小浜町雲仙316
  0957-73-3255
- **島原温泉 ホテル南風楼**
  島原市弁天町2-7331-1
  0957-62-5111
- **国民宿舎 青雲荘**
  雲仙市小浜町雲仙500-1
  0957-73-3273
- **HOTEL シーサイド島原**
  島原市新湊1-38-1
  0957-64-2000

※最新の販売店情報はホームページでご確認ください。
Please check our website for the latest retail shop information.

◆**他にも素敵な場所があります。あなただけの旅の宝を探してみましょう!**
**There are other wonderful places to visit. Let's take a trip to find a treasure of your own!**

20

## HIRAIZUMI・HANAMAKI -Iwate-

# 平泉・花巻 (岩手県)

岩手県南部に古くから存在する観光地である平泉。東北有数の温泉地・花巻。平安時代末期から残る寺院や遺跡など、歴史的にも価値がある地域で世界文化遺産にも登録されている観光名所です。

It is a tourist resort which exists for a long time located in the southern part of Iwate prefecture. There are temples and ruins that exist from the end of Heian period, and it is a very worthy place historically; many places are registered as the World Heritage Site.

## 旅 TABI-SAKE 酒

### HIRAIZUMI・HANAMAKI

岩手県産米を100%使用し、清らかな穀物香に木の実や笹の葉などに似た香りが混じり合いながら透明感のある香りを織りなしています。きめ細やかで爽快な酸味と旨みとのバランスが絶妙です。

Made from 100% rice from Iwate prefecture, this Junmai Ginjo-shu has a clear aroma derived from a fresh grain-like scent, mixed with nut and bamboo-like fragrances. The balance between fine, refreshing sourness and mellow taste is superb.

**醸造元**：株式会社桜顔酒造
(1952年創業／岩手県)
**種類**：日本酒 純米吟醸酒
**度数**：15度
**精米歩合**：55%

**Brewery** : Founded Ltd. Sakuragao brewing (Since 1952／Iwate)
**Kind** : Junmai Ginjo-shu
**Alcohol Content** : 15%
**Polishing Ratio** : 55%

甘辛度
Sweet-dryness

## 主な行き方 Major Directions

羽田空港＝(20分)⇒浜松町駅＝(5分)⇒東京駅＝(115分)⇒一ノ関駅＝(10分)⇒平泉駅＝(40分)⇒花巻駅

Haneda Airport＝(20 min)⇒Hamamatsucho Station＝(5 min)⇒Tokyo Station＝(115 min)⇒Ichinoseki Station＝(10 min)⇒Hiraizumi Station＝(40 min)⇒Hanamaki Station

### ▶①中尊寺

世界遺産に登録されている中尊寺は、859（嘉祥3）年に比叡山延暦寺の高僧慈覚大師円仁によって開かれました。中尊寺には、非戦の決意と平和への願いが込められています。中尊寺の堂塔のなかでも、特に金色堂は創建当時の姿を残す唯一の建造物です。敵味方関係なく極楽浄土に行けるようにという願いが込められているその姿は、金箔が輝く緻密で繊細な作りになっています。

### ▶①Chusonji Temple

Registered as the World's Heritage, the Chusonji Temple was built in 859 by a high priest of Tendai Sect: Jikaku Daishi Ennin. The Chusonji Temple has wishes of no-war and peace. Out of all the temple buildings, especially the Kojikidou has its features left from when it was built. It has wishes that everyone can go to heaven, and it is built delicately with gold foil shining on the outside.

### ▶②宮沢賢治記念館

日本を代表する詩人であり、作家である、宮沢賢治の暮らした花巻の町には、賢治作品の題材やモデルとなった場所、賢治にまつわるスポットがたくさんあります。多彩なジャンルに及ぶ宮沢賢治の世界と出会えるこの施設は、年間20万人を超える訪問客を迎えています。

### ▶②Miyazawa Kenji Memorial Museum

Hanamaki city is where Kenji Miyazawa, a Japanese famous poet and children's story writer, lived, and it is the model of many works made by him in the city. The memorial house has many different genres of Kenji Miyazawa's artistic world, and over 200,00 people visit here in a year.

**Brewery's recommended spot**
**蔵元おすすめスポット**

## えさし藤原の郷

奥州市江刺区岩谷堂小名丸86-1
86-1, Konamaru, Iwayado, Esashi-ku, Oshu-shi

約20ヘクタールの敷地に日本で唯一、平安貴族の住宅、寝殿造の建物を再現した「伽羅御所」など、大小約120棟の建物が建てられている。

**Esashi-Fujiwara Heritage Park**
In the area, about 20 hectares, there is the house of the Heian period aristcrats, a mock of "Kyaranogosho," and 120 many other big and small buildings.

## 旅酒が見つかる場所 Shop

● **ホテル千秋閣**
花巻市湯本1-125
0198-37-2150

● **ホテル花巻**
花巻市湯本1-125
0198-37-2180

● **ホテル紅葉館**
花巻市湯本1-125
0198-37-2140

● **石鳥谷観光物産館 酒匠館**
花巻市石鳥谷町中寺林
第7地割17-3
0198-45-6868

● **ホテル志戸平**
花巻市湯口字志戸平26
0198-25-2011

※最新の販売店情報はホームページでご確認ください。
Please check our website for the latest retail shop information.

◆**他にも素敵な場所があります。あなただけの旅の宝を探してみましょう!**
**There are other wonderful places to visit. Let's take a trip to find a treasure of your own!**

(佐賀県)

「虹の松原」を始めとする恵まれた自然とそこでとれる食材、「唐津城」などの歴史的な遺産や「唐津焼」のような名産品。唐津には多くの魅力が詰まっています。

Starting with "The rainbow of Matsubara," rich nature and great ingredients collected there, historical inheritance such as the Karatsu castle, and great food such as the Karatsu-yaki (Karatsu Ware). There are many attractions in Karatsu.

## 旅酒 TABI-SAKE KARATSU

佐賀県産「佐賀の華」を55%精米歩合で仕込んだ純米吟醸酒、華やかな味わい深いうまさが特徴です。幅広いシーンに適しています。料理との相性も良く、冷やして冷や、ぬる燗(10〜40℃)がおすすめです。

The Junmai Ginjo-shu uses 55% of "Saganohana," from Saga, and has a gorgeous and deep taste. It is suited for all types of scenes. It goes well with food and is great when it's cold or at 10-40°C.

**醸造元**:窓乃梅酒造株式会社
(1688年創業/佐賀県)
**種類**:日本酒 純米吟醸酒
**度数**:15度
**精米歩合**:55%

**Brewery** : Madonoume Shuzo Co., Ltd. (Since 1688/Saga)
**Kind** : Junmai Ginjo-shu
**Alcohol Content** : 15%
**Polishing Ratio** : 55%

甘辛度
Sweet-dryness

## 主な行き方 Major Directions

羽田空港＝(120分)⇒福岡空港＝(90分)⇒唐津
Haneda Airport＝(120 min)⇒Fukuoka Airport＝(90 min)⇒Karatsu Station

## 旅の宝 Treasure Hunt

### ▶①浜野浦の棚田

玄海町・浜野浦地区の棚田は、日本棚田百選にも選ばれています。283枚の田が連なる11.5ヘクタールの棚田は展望台から見下ろすことができます。特に田植えの時期の夕日が沈む時間には、太陽の光が反射しながら輝く田んぼを眼下いっぱいに見ることができ、その美しさに心を奪われることでしょう。

### ▶①The Tanada (Rice Terraces) of Hamanoura

The Tanada in the Genkai Town and the area of Hamanoura is selected as the 100 best Tanadas in Japan. People can look down at the view of 283 rice fields making 11.5 hectares of Tanada from an observation platform. Especially when the sun falls during the seasons of rice planting, the light reflects on the field and shines all over the place.

### ▶②唐津城

別名「舞鶴城」とも呼ばれる唐津城。天守閣を鶴の頭、左右の松原を翼に見立て、その名が付いています。天守閣からは、日本三大松原のひとつである「虹ノ松原」を望むことができるために眺望のよさでも知られており、春には桜の名所としても知られています。眺望だけではなく、掘割、舟運といった築城技術による景観のすばらしさも見所となっています。

### ▶②Karatsu Castle

Karatsu Castle is also called as the Maizuru Castle. The name comes from the crane's head on top of the castle tower and the wings from the pine covered area at the left and right side of the castle. From the castle tower, you can see the pine covered area, which is the 3 largest in Japan. Also, in spring it is famous for cherry blossoms. Not only the view but also the building techniques are great points to see.

## Brewery's recommended spot
## 蔵元おすすめスポット

### 唐津くんち

唐津市南城内3-13
3-13, Minamijonai, Karatsu-shi

50万人が訪れる活気あふれる町人文化。1800年代から作られた数々の曳山あり。唐津神社の秋季例大祭で唐津最大の行事。国の無形民俗文化財に指定。

### Karatsu Kunchi

About 500,000 people visit this townsmen culture. There are many festival floats made in the 1800s and the biggest festival of Karatsu is held here. It is the country's Intangible Folk Cultural Property.

### 旧高取家住宅

唐津市北城内5-40
5-40, Kitajonai, Karatsu-shi

炭鉱主・高取伊好の邸宅で唐津城本丸近くの海沿いに約2300坪の敷地に約300坪の2棟の建物。公と私で区分され、公の部分には能舞台が現存（国内唯一）。

### Former Takatori House

Coal mine holder. A house of Koreyoshi Takatori and is near the Karatsu Castle along the ocean. It has about 7600m² in area and there are 2 buildings that are about 1000m². It is divided into two uses which are for public and own use. The public use building has a theater for Noh.

## 旅酒が見つかる場所 Shop

- **HOTEL&RESORTS SAGA-KARATSU**
  唐津市東唐津4-9-20
  0955-72-0111
- **マリンセンター おさかな村**
  唐津市浜玉町浜崎1922
  0955-56-2200
- **日本料理 旅館 水野**
  唐津市東城内4-50
  0955-72-6201
- **道の駅 桃山天下市**
  唐津市鎮西町名護屋1859
  0955-51-1051
- **竹八**
  佐賀市駅前中央1-11-1
  えきマチ1丁目内
  0952-24-0777

※最新の販売店情報はホームページでご確認ください。
Please check our website for the latest retail shop information.

22

## HAGI -Yamaguchi-

# 萩 （山口県）

歴史的な街である山口・萩。幕末から明治にかけて、日本を担った人材を数多く輩出した松下村塾をはじめ、観光名所が多い街です。自然豊かで美味しい食材も豊富。歴史に触れ合いながら、さまざまな萩の魅力を堪能してください。

Hagi, Yamaguchi prefecture, is a historical city. From the end of Edo period to the Meiji period, the Shokason-juku School discharged many people who bore the future of Japan, and many other tourist resorts exist. The nature is rich and the ingredients are abundant. Please enjoy the attractions of Hagi along with touching the history of Hagi.

## 旅 TABI-SAKE 酒

### HAGI

県独自の酒米「西都の雫」を使った純米吟醸酒です。華やかで豊かな香りと上品な甘みを持った飲み口爽やかなお酒です。

It is the Junmai Ginjo-shu using prefecture own sake rice“Saitono shizuku.” It is colorful and rich aroma and elegant sweetness have been drinking a refreshing drink.

**醸造元**：岩崎酒造株式会社
（1901年創業／山口県）
**種類**：日本酒 純米吟醸酒
**度数**：16度以上17度未満
**精米歩合**：40%

**Brewery** : Iwasaki Shuzo Co., Ltd.
(Since 1901／Yamaguchi)
**Kind** : Junmai Ginjo-shu
**Alcohol Content** : 16%-17%
**Polishing Ratio** : 40%

甘辛度
Sweet-dryness
甘 辛
Sweet Dry

## 主な行き方 Major Directions

羽田空港＝(95分)⇒山口宇部空港／バス＝(70分)⇒萩市内
Haneda Airport＝(95 min)⇒Yamaguchi-ube Airport／Bus＝(70 min)⇒Hagi-shi

羽田空港＝(80分)⇒萩・石見空港／バス＝(15分)⇒益田駅＝(80分)⇒萩駅
Haneda Airport＝(80 min)⇒Hagi･Iwami Airport／Bus＝(15 min)⇒Masuda Station＝(80 min)⇒Hagi Station

## 旅の宝 Treasure Hunt

### ▶①萩城下町

萩は日本有数の城下町です。江戸時代の街割や町屋などがそのまま現存している部分も多く、古地図をみながらでも歩ける町として知られています。街は平坦で歩きやすい場所が多く、歴史を感じることのできる建造物が密集しているため、街歩きにはぴったりです。武家屋敷や土壁など城下町としての面影が今尚色濃く残っており、タイムスリップしているかのような気分を味わうことができるでしょう。

### ▶②松下村塾

松下村塾は、吉田松陰が主宰していた塾として観光では人気のスポットとなっています。吉田松陰は身分や階級にかかわらず、塾生を多く受け入れ、伊藤博文や高杉晋作などを育てたことでも有名です。松下村塾は、そんな吉田松陰の歴史と栄光をたたえ、修復を重ねながらも当時の姿を残しています。

### ▶①Hagi Castle town

Hagi is one of the many famous castle towns. There are still shops and streets that exist from the Edo period and it is famous for being able to walk the town by looking at an old map. The city is flat and many historical buildings are gathered at one place so it is easy to walk around the city. The old samurai residence and the dirt walls are remaining features from the past, so people can feel as if they have transported through time.

### ▶②Shokason-juku School

Shokason-juku School is a famous tourist spot for Shoin Yoshida being the president. Shoin Yoshida in took many students without segregation, and is famous for educating Hirobumi Ito and Shinsaku Takasugi. Shokason-juku School has its features remaining from that time by repairing and is praising the history and glory of Shoin Yoshida.

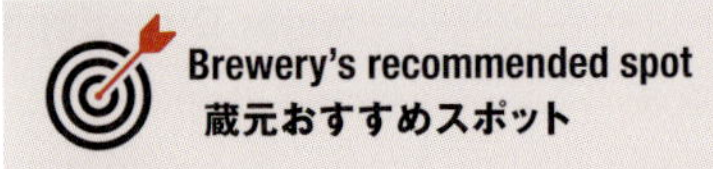

## Brewery's recommended spot
## 蔵元おすすめスポット

### 明倫学舎

萩市江向602
602, Emukai, Hagi-shi

全国屈指の規模を誇った萩藩校明倫館の跡地に建つ日本最大の木造校舎。2014年3月まで授業が行われていた萩の新しい観光起点である。

### Meirin Gakusha (School House)

The Meirin Gakusha, the largest wooden building in Japan, is still built on the remained site of the Hagi han school Meirinkan. It had classes held till March 2014 and is a new tourist resort.

### 萩の「明治日本の産業革命遺産」

### Sites of Japan's Meiji Industrial Revolution in Hagi

Hagi Castle town and Shokason-juku School are part of Sites of Japan's Meiji Industrial Revolution and are registered as the World' Heritage Site in 2015. In Hagi, there are also three Sites of Japan's Meiji Industrial Revolution: Hagi Reverberatory Furnace, Ebisugahana Shipyard and Oitayama Tatara Iron Works.

萩城下町・松下村塾は、2015年に「明治日本の産業革命遺産」として世界遺産に登録されている。その他「萩反射炉」「恵美須ヶ鼻造船所跡」「大板山たたら製鉄遺跡」も萩エリアの世界遺産である。

## 旅酒が見つかる場所 Shop

- **城跡ながお**
  萩市堀内83-27
  0838-25-3892
- **松陰神社 売店**
  萩市椿東1537 松陰神社境内
  0838-22-1800
- **萩本陣 売店**
  萩市椿東385-8
  0838-25-0385
- **萩の宿 常茂恵**
  萩市土原弘法寺608-53
  0838-22-0150

※最新の販売店情報はホームページでご確認ください。
Please check our website for the latest retail shop information.

## ASUKA -Nara-

# 飛鳥 (奈良県)

日本創世の魅力を味わうことができる飛鳥。多くの古墳や遺跡が残る歴史好きにはたまらない街です。歴史名所を巡りつつ、古代の日本に思いを馳せてはいかがでしょうか。

Aska is a place where you can enjoy the genuine of Japan. Many ancient tombs and ruins are remaning. Why don't you walk around and feel the ancient Japan?

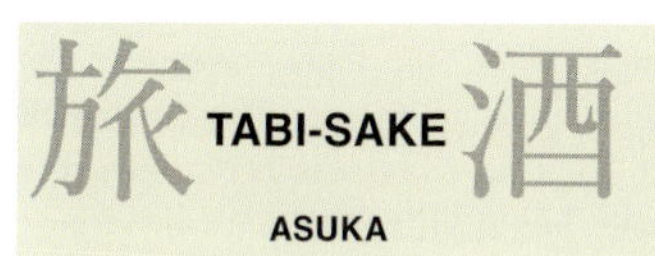

## 旅 TABI-SAKE 酒

ASUKA

調和のとれた淡麗辛口。キレのよい軽快な口あたりは飲むほどに冴え渡ります。

Having a harmony of dry and crispy taste. The texture is light and refreshing, making it more distinctive when you drink more.

**醸造元：**喜多酒造株式会社
（1718年創業／奈良県）
**種類：**日本酒 純米酒
**度数：**15度
**精米歩合：**麹米65%、掛米70%

**Brewery** : Kita Shuzo Co., Ltd.
(Since 1718／Nara)
**Kind** : Junmai-shu
**Alcohol Content** : 15%
**Polishing Ratio** : koji rice 65%、kake rice 70%

甘辛度
Sweet-dryness

## 主な行き方 Major Directions

羽田空港＝(65分)⇒大阪国際空港／バス＝(40分)⇒大阪阿部野橋駅＝(35分)⇒橿原神宮前駅＝(5分)⇒飛鳥駅

Haneda Airport＝(65 min)⇒Osaka International Airport／Bus＝(40 min)⇒Osakaabenobashi Station＝(35 min)⇒Kashihara Jingumae Station＝(5 min)⇒Asuka Station

### ▶①飛鳥寺

596年に蘇我馬子によって建てられた日本最古の寺院であり、現在は安居院と呼ばれています。飛鳥寺のなかでは、日本最古の仏像である飛鳥大仏をみることができます。本堂では住職の方による説明を受けたり、展示室では土器や仏像をみることができるなど、飛鳥の時代を今に伝える貴重な寺院となっています。

### ▶②石舞台古墳

総重量2300トンの岩を使っているとされる日本最大級の古墳です。飛鳥時代の名所を巡る観光スポットとして代表的ではありますが、その被葬者は明らかになっていません。蘇我馬子という説が有力ですが、蘇我稲目という説もあり、いまだに謎に包まれているということも魅力の一つになっているのかもしれません。また、周囲には桜の木が植えられており、春にはライトアップされた石舞台古墳と夜桜を楽しむことができます。

### ▶①Asukadera Temple

In 596, the Asukadera Temple, known as Angoin today, was built by Soga no Umako and is the oldest temple in Japan. In the Asukadera temple, you can also see the Asuka Daibutsu. In the main building, the workers will explain you the informations, and in the exhibition roon, you can see the crafts made by dirt. It is a very important temple to descend the atmosphere of the Asuka period.

### ▶②Ishibutai Tomb

The tomb is the largest one in Japan, using about 2300 tons of stones. It is a famous spot to tour; however, the entombed is unknown. The statement that Soga no Umako is being buried is strong, but is still filled with mysteries. Furthermore, there are cherry blossoms around the tomb, and in spring the trees are lighted up at night. You can enjoy the night cherry blossoms with the tomb.

## Brewery's recommended spot
## 蔵元おすすめスポット

### 橿原神宮

橿原市久米町934
934, Kume-cho, Kashihara-shi

初代天皇「神武天皇」が即位した「日本のはじまりの地」とされる。
本殿と文華殿は重要文化財に指定されている。

### Kashiharajingu Shrine

It is the place where the first emperor was enthroned in Japan is the place where Japan started. The main building and the Bunkaden are both the country's Important Cultural Properties.

## 旅酒が見つかる場所 Shop

- **かしはらナビプラザ 物産コーナー**
  橿原市内膳町1-6-8
  0744-47-2270
- **なら和み館**
  奈良市高畑町1071
  0742-21-7530

※最新の販売店情報はホームページでご確認ください。
Please check our website for the latest retail shop information.

◆**他にも素敵な場所があります。あなただけの旅の宝を探してみましょう!**
**There are other wonderful places to visit. Let's take a trip to find a treasure of your own!**

## AZUMINO・SHIRAHONE ONSEN -Nagano-

# 24 安曇野・白骨温泉（長野県）

雄大な北アルプスと自然豊かな田園地帯が広がる安曇野エリア。自然豊かなこのエリアでは、豊富な山の幸とどこか懐かしい日本の田園風景に出会えます。また、白骨温泉は鎌倉時代から続く信州の秘湯。大自然に囲まれた環境は、季節ごとに違った表情をみせ、多くの観光客が訪れます。周囲の名所はその大自然の魅力を堪能できるものばかり。温泉と合わせて観光名所をゆっくりと巡ってみてはいかがでしょうか。

The Azumino area has the brilliant north alps and the rich nature. In this area, you can enjoy the food from the mountains and the view of the rural scene of Japan. The Shirahone Onsen is the secret spa of the shinshu area and has its history from the Kamakura period. There are famous places where you can enjoy the nature around the spa. Why don't you visit the spots with the spa?

### 旅 TABI-SAKE 酒

**AZUMINO・SHIRAHONE ONSEN**

お米の味わいと吟醸酒らしいフルーティーな味わいの両方を楽しめます。冷やすと味が締まり、さらに美味しく召し上がっていただけます。長野県産米美山錦100%使用。どんな料理にも合います。

You can both enjoy the taste of rice and fruitiness. When it is cold the taste is tightened, and it becomes even better. It uses 100% Miyamanishiki from Nagano. It goes well with any type of food.

**醸造元**：笑亀酒造株式会社
（1883年創業／長野県）
**種類**：日本酒 純米吟醸酒
**度数**：15度
**精米歩合**：59%

**Brewery** : Shoki Shuzo Co., Ltd.
（Since 1883／Nagano）
**Kind** : Junmai Ginjo-shu
**Alcohol Content** : 15%
**Polishing Ratio** : 59%

**甘辛度**
Sweet-dryness

## 主な行き方 Major Directions

羽田空港＝（20分）⇒品川駅＝（20分）⇒新宿駅＝（170分）⇒松本駅／バス＝（130分）⇒白骨温泉

Haneda Airport＝(20 min)⇒Shinagawa Station＝(20 min)⇒Shinjuku Station＝(170 min)⇒Matsumoto Station／Bus＝(130 min)⇒Shirahone Onsen

## 旅の宝 Treasure Hunt

### ▶①上高地

国の特別名勝にも指定されている有数の景勝地である上高地。4月下旬から11月中旬まで開山しており、多くの観光客・登山客を集めています。一面に咲くニリンソウの花や、大正池の透明な水面に移る穂高連峰、黄色い色に染まっていくハルニレやカツラなど、日々変わる景色は何度足を運んでも飽きることがないでしょう。

### ▶②乗鞍三名滝

日本の屋根とも呼ばれている北アルプス大自然の恩敬を最大限に受けている乗鞍高原。そこには乗鞍三名滝と呼ばれる「三本滝」「善五朗の滝」「番所大滝」という名所があります。それぞれが異なる標高に位置するため、水量や落差など違った姿を見せてくれます。

### ▶③穂高神社

穂高駅から徒歩3分のところに位置する神社。地元では交通安全の神様として知られ、交通安全祈願に訪れる人が多くみられます。

### ▶①Japan Alps Kamikochi

Japan Alps Kamikochi is designated as the country's Special Places of Scenic Beauty. The mountain is opened from late April to middle of November, and many tourists and mountain climbers visit. The bloom of windflower all over the mountain, the reflection of the mountains on the clear water surface of Lake Taisho, the yellow kastura trees, and many other views changing daily would never make tourists bored.

### ▶②Three Major Waterfalls of Mt. Norikura

The Norikura Waterfall is structured by the Japanese Alps, also called as the roofs of Japan. At the Norikura Waterfall, there are three famous ones: Sanbontaki Waterfall, Zengoro's Waterfall, and Bansho-Oodaki Waterfall. Each waterfall is different in its height, thus, making each feature different.

### ▶③Hotaka Shrine

A shrine located 3 min walk from Hotaka Station. It is known for traffic safety from the local people, and many people visit here to pray.

## Brewery's recommended spot
## 蔵元おすすめスポット

### 阿礼神社例大祭

塩尻市塩尻町6

6, Shiojiri-machi, Shiojiri-shi

別名塩尻祭り。七地区（上町・室町・中町・宮本町・堀の内・長畝・桟敷）で構成された氏子から山車が出て阿礼神社に集結する姿は必見。

### Annual Grand Festival of Areijinja Shrine

It is called as the Shiojiri-Matsuri. The Ujiko (shrine parishioner) made from 7 different areas gathering at the shrine is a must see festival.

### 大王わさび農場

安曇野市穂高3640

3640, Hotaka, Azumino-shi

日本一広いわさび田を持つ農場。安曇野観光では定番のスポット。水車小屋のある風情ある風景でも有名。わさびソフトクリームや、わさびを使った料理も多くあり、見学だけでなく、食事もできる。

### Daio Wasabi Farm

A farm that has the largest wasabi farm in Japan. A famous place to visit in Azumino. It is also famous for having a water mill. not only seeing the farm, but also enjoying the foods using wasabi such as wasabi softcream is a great way to enjoy.

## 旅酒が見つかる場所 Shop

- **中島屋**
  松本市安曇706-1
  0263-94-2007
- **白船グランドホテル**
  松本市安曇4203
  0263-93-3333
- **泡の湯旅館**
  松本市安曇4181
  0263-93-2101

※最新の販売店情報はホームページでご確認ください。
Please check our website for the latest retail shop information.

# TAKACHIHO -Miyazaki-

25

# 高千穂（宮崎県）

緑の山々に囲まれた日本神話の舞台、高千穂。神々を奉る神社をはじめとした、神秘的な名所が数多く存在し、天然記念物や重要文化財に指定されています。悠久の自然に囲まれた神々ゆかりの地を、ごゆっくりお楽しみください。

Takachiho is surrounded by green and the stage for Japanese mythology. Many shrines where the gods live and many mysterious places exist and many places are designated as the Natural Monuments and Important Cultural Properties. Please enjoy the land surrounded by the nature and associated with gods.

## 旅酒 TABI-SAKE TAKACHIHO

30有余年で培った巧みなブレンド。貯蔵技術により芳醇な香りとまろやかな味わいを醸し出します。氷点下20℃でろ過することによって渋みの原因となる成分を軽減し、はっきりとしたキレの良い味わいと芳香が引き出されています。麦のコクと樽の風味がほどよく融合した厚みのある口あたりは、熟成感の余韻がしっかり残る風味となっています。

The blend of sake cultivated in 30 years. The technique of storage has made the shochu have a rich bouquet and a soft taste. By filtering it at -20℃, the astringent taste disappears, and the taste becomes refreshing. The deep taste of wheat and the scent of the barrel mixes up well in your mouth, having a maturing aged shochu flavor.

**醸造元**：神楽酒造株式会社
（1954年創業／宮崎県）
**種類**：本格焼酎 長期貯蔵酒
**度数**：25度

**Brewery** : Kagura Brewing Co., Ltd.
（Since 1954／Miyazaki）
**Kind** : Honkaku-shochu（Long-term Storage Liquor）
**Alcohol Content** : 25%

主な行き方 Major Directions

羽田空港＝（105分）⇒宮崎空港＝（60分）⇒延岡駅／バス＝（60分）⇒高千穂駅

Haneda Airport＝(105 min)⇒Miyazaki Airport＝(80 min)⇒Nobeoka Station／Bus＝(80 min)⇒Takachiho Station

### ▶①高千穂峡

太古の昔、阿蘇山の火山活動で噴出した火砕流が侵食されて柱状節理の素晴らしい断崖となった峡谷で、1934年に国の名勝・天然記念物に指定されています。遊歩道が整備され、日本の滝百選に選ばれた真名井の滝をはじめ、柱状節理の峡谷美が楽しめます。

### ▶①Takachiho Gorge

In the past, due to the volcanic activity of Aso mountain, the pyroclastic flow being erupted has eroded the soil to an unique shape and the gorge is designated as the Natural Monument in 1934. There is also a bridge built so that people can see the surface of the gorge along with the Manainotaki waterfall.

### ▶②天岩戸神社

日本神話（古事記・日本書紀）に書かれている天岩戸神話の伝承地となっている場所。天照大神がお隠れになった天岩戸と呼ばれる洞窟を御神体としてお祀りする西本宮と、天照大神をお祀りする東本宮があります。

### ▶②Amanoiwatojinja Shrine

It is a famous place that is written in the Japanese mythology (*Kojiki* and *Nihon Shoki*), and the cave where the Amaterasu Omikami is hiding is called the Amanoiwato. There are two places where the god is worshiped the westhongu and the east hongu.

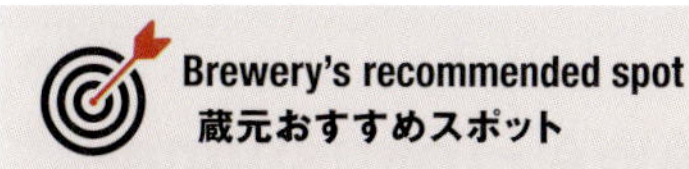

## Brewery's recommended spot 蔵元おすすめスポット

### 常光寺の滝

西臼杵郡高千穂町岩戸

Iwato, Takachiho-cho, Nishiusuki-gun

天岩戸神社から北へ向かった「上岩戸」という辺りにある常光寺の滝。別名「白糸の滝」とも言われ、大きい滝だけではなく、二段目の滝もとても美しい。
天岩戸神社から向かうと、15〜20分くらいで行けるが、狭い山道であり運転にはかなりの注意が必要。しかし、この滝を見た瞬間に到着するまでの苦労は忘れてしまうだろう。

### Jokojinotaki Waterfall

Going north from the Amanoiwatojinja Shrine, there is “Kamiiwato”, where the warterfall is. It is also called as the Shiroitonotaki Waterfall and is very beautiful. It is not only big but also the second level of the waterfall is a small scale but beautiful. If you are heading from Amanoiwatojinja Shrine, there are many marks to lead you to the waterfall: however, the street is very narrow, so you have to be careful while driving. It is very tiring to get to the fall but once you get there, you will forget all the hard work to get there.

### 国見ヶ丘

西臼杵郡高千穂町押方

Oshikata, Takachiho-cho, Nishiusuki-gun

その昔、神武天皇の御孫・建磐竜命（たていわたつのみこと）が九州統制の折、この丘に立って国見されたことから「国見ヶ丘」と呼ばれるようになったと伝えられている。晴天の日には、高千穂盆地や遠く阿蘇の山々を望むことができ、10月上旬から11月下旬の早朝には、雲の海に山々が島のように浮かんで見える雲海を見ることができる。

### Kunimigaoka Hill

It is called “Kuminigaoka” because the Tateiwatatsunomikoto, the grandson of Jimmu emperor, saw this country from a hill. On a sunny day, you can see the Takachiho basin and Mt. Aso, during early October to late November, you can see the tip of the mountains floating on the sky like islands.

## 旅酒が見つかる場所 Shop

- **トンネルの駅**
  西臼杵郡高千穂町下野2221-2
  0982-73-4050
- **千穂の家**
  西臼杵郡高千穂町向山62-1
  0982-72-2115
- **まちなか案内所**
  高千穂町大字三田井802-3
  0982-72-3031
- **みやげ店 安河原**
  西臼杵郡高千穂町大字岩戸1082
  090-5725-7083
- **野崎漬物株式会社 JR店**
  宮崎市錦町1-8 宮崎駅構内
  0985-27-3955

※最新の販売店情報はホームページでご確認ください。
Please check our website for the latest retail shop information.

26

## YOSHINOGAWA -Tokushima-

# 吉野川（徳島県）

四国最大の河川、吉野川に育まれたこの地域は、四季折々の自然を楽しむことができます。流域には、渓谷で有名な「大歩危・小歩危」や「美濃田の渕」があり、素晴らしい景色を楽しむことも。ラフティングの名所としても有名で、観光からアウトドア、さまざまな魅力を味わってはいかがでしょうか。

The area being natured by the largest river in Shikoku, Yoshino River, is famous for the beauty of nature. At the basin, there are many famous valleys such as the Oboke, Koboke and Minodanohuchi. It is also famous for rafting and many other outdoor sports. Why don't you enjoy the variety of attractions of Yoshino River?

## 旅 TABI-SAKE 酒

### YOSHINOGAWA

美郷産鶯宿梅の力強さを前面に押し出した梅酒。ロック推奨。

Plum wine pushed out the powerfulness of Misato Oushuku plum to the front. It is recommended to drink it on the rocks.

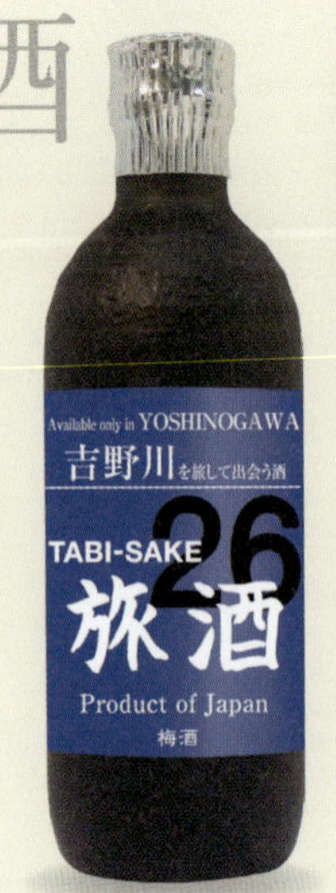

**醸造元**：徳長梅酒製造場
（2015年創業／徳島県）
**種類**：梅酒
**度数**：16度

**Brewery** : Tokunaga Plum Winery
（Since 2015／Tokushima）
**Kind** : Plum Wine
**Alcohol Content** : 16%

Major Directions

羽田空港＝(75分)⇒徳島空港／バス＝(30分)⇒徳島駅＝(30分)⇒鴨島駅

Haneda Airport＝(75 min)⇒Tokushima Airport／Bus＝(30 min)⇒Tokushima Station＝(30 min)⇒Kamojima Station

旅の宝 Treasure Hunt

### ▶①美郷ほたる館

1970年に「ホタルおよびその発生地」として、国の天然記念物に指定された美郷。ホタルが飛ぶ流域面積、数、期間とも全国でも有数の地域となるこの一帯。ほたるに関する資料展示や研究、飼育を行っている施設です。

### ▶①Misato Hotaru (Firefly) Museum

In 1970, Misato was registered as the country's natural monument due to the amount of fireflies. This place is one of the few places where fireflies fly around and there are many exhibitions and studies conducted about fireflies.

### ▶②川島城

吉野川沿いの中世山城。阿波九城のひとつでもあり、室町時代からある歴史ある名所です。かつての本丸跡からは吉野川を一望することができ、多くの観光客が訪れる観光スポットとなっています。

### ▶②Kawashima Castle

It is along the Yoshino River. It is standing from the Muromachi era and is very famous. From the honmaru, you can see the whole Yoshino River and many tourists visit here to see the beauty of the river.

## Brewery's recommended spot
## 蔵元おすすめスポット

### 美郷梅酒祭り

吉野川市美郷宗田82-1

82-1, Misatomuneda, Yoshinogawa-shi

毎年11月の最後の土日に開催している。無料シャトルバスで美郷をめぐりながら、梅酒を味わうイベント。

### Misato Ume-shu Festival

It is held on the last week of Saturday and Sunday of November every year. It is an event that you ride a free shuttle bus and visit around Misato and taste Plum Wine.

## 旅酒が見つかる場所 Shop

- **美郷物産館**
  吉野川市美郷字峠463-3
  0883-26-7888
- **かずら橋夢舞台**
  三好市西祖谷山村今久保345-1
  0883-87-2200

※最新の販売店情報はホームページでご確認ください。
Please check our website for the latest retail shop information.

◆**他にも素敵な場所があります。あなただけの旅の宝を探してみましょう!**
**There are other wonderful places to visit. Let's take a trip to find a treasure of your own!**

## KATSURAHAMA -Kochi-

# 27 桂浜（高知県）

月の名所として古くから有名な桂浜。「よさこい節」でもその名が詠われており、高知県の代表的な観光名所となっています。桂浜周辺では、坂本龍馬像が置かれている桂浜公園でリラックスしたり、坂本龍馬記念館や桂浜水族館などの施設を訪れたり、さまざまな楽しみ方ができます。

As famous for the moon, Katsurahama is famous for its long time ago. Even in “Yosakoi clause” its name is recited and it is becoming a representative tourist attraction in Kochi prefecture. In the vicinity of Katsurahama, you can relax at the Katsurahama Park where Sakamoto Ryoma statue is located and enjoy various facilities such as Sakamoto Ryoma Memorial Museum and Katsurahama Aquarium.

## 旅 TABI-SAKE 酒

**KATSURAHAMA**

品の良いナチュラルな香りとなめらかにふくらむ味わい、そしてあと口は抜群のキレを誇り、そのバランスの良さは食中酒としての完成度の高さを表しています。

It has an elegant scent and a soft taste, the after taste is dry and refreshing which makes it go well with any type of food.

**醸造元：**司牡丹酒造株式会社
（1603年創業／高知県）
**種類：**日本酒 特別純米酒
**度数：**15度以上16度未満
**精米歩合：**60%

**Brewery** : TSUKASABOTAN Brewing Co., Ltd.（Since 1603／Kochi）
**Kind** : Special Junmai-shu
**Alcohol Content** : 15-16%
**Polishing Ratio** : 60%

**甘辛度**
Sweet-dryness
甘 Sweet —— 辛 Dry

## 主な行き方 Major Directions

羽田空港＝(75分)⇒高知空港／バス＝(25分)⇒高知駅＝(15分)⇒栈橋通五丁目／バス＝(20分)⇒龍馬記念館前／徒歩＝(5分)⇒桂浜

Haneda Airport＝(75 min)⇒Kochi Airport／Bus＝(25 min)⇒Kochi Station＝(15 min)⇒Sanbashidori5chome／Bus＝(20 min)⇒Ryoma Kinenkanmae／Walking＝(5 min)⇒Katsurahama

## 旅の宝 Treasure Hunt

### ▶①高知城

土佐藩の初代藩主山内一豊によって造られた高知城は、現在は国の史跡に指定され、高知県のシンボル・観光名所になっています。天守や御殿などは重要文化財にも指定されており、本丸が現存している唯一の城であることなどから、多くの観光客が訪れています。

### ▶①Kochi Castle

Kochi Castle, built by Kazutoyo Yamanouchi, the first lord of the Tosa han, is now designated as a national Historic Site, becoming a symbol and tourist attraction of Kochi prefecture. Many tourists are visiting since the castle towers and the Gotens are also designated as Important Cultural Properties, as Honmaru is the only castle in existence.

### ▶②坂本龍馬像

桂浜に置かれた坂本龍馬像は、太平洋を見つめるようにして立っています。全国にある坂本龍馬像のなかでも一番高い銅像であり、全長は13m以上あります。自由に見学することができる、市民の憩いの場でもあります。

### ▶②Sakamoto Ryoma Bronze Statue

Sakamoto Ryoma Bronze Statue at Katsurahama is standing to look at the Pacific Ocean. It is the highest bronze statue among the Sakamoto Ryoma statues in the whole country, and the total length is 13 m or more. It is also a place for citizens to relax, where visitors can freely visit.

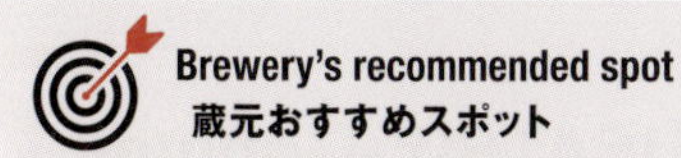


## 仁淀川

日本一水のきれいな川。仁淀ブルーと呼ばれ、近年、写真家や観光通の間でブームになっている。

## Niyodo River

It is a river that has the most beautiful water in Japan. It is called Niyodo Blue and recently many photographers and families are visiting here.

## 旅酒が見つかる場所 Shop

- **黒岩観光みやげセンター**
  高知市浦戸779-2
  088-842-2737
- **土佐の地酒市場 西寅**
  高知市帯屋町2-3-1
  ひろめ市場内
  088-872-6676

※最新の販売店情報はホームページでご確認ください。
Please check our website for the latest retail shop information.

**◆他にも素敵な場所があります。あなただけの旅の宝を探してみましょう!**
**There are other wonderful places to visit. Let's take a trip to find a treasure of your own!**

# 28

## KUSATSU -Gunma-

（群馬県）

日本を代表する名泉のひとつ、草津。古くから湯治の場として親しまれてきました。日本三名泉に上げられる草津の温泉を満喫しながら、観光をお楽しみください。また、標高1200mの高原に位置する草津温泉では、温泉以外にも豊かな自然に囲まれているため、四季折々の景観を楽しむことができます。

Kusatsu is one of the famous spa resorts in Japan. It is popular among people for medical purposes. Please enjoy the trip by enjoying the one of the top three spas in Japan. Located on a plateau 1,200 meters above sea level, Kusatsu Onsen is surrounded by nature, where you can enjoy the seasonal scenery as well as hot spring bathing.

### 旅 TABI-SAKE 酒

**KUSATSU**

米の旨みを引き出すべく、米と水のみを原料として、雪滴水で醸し、芳醇な香りとのどごしが良く、飲みやすいお酒です。

In order to pull out the taste of umami of the rice, it only uses rice and water for its ingredients, and it has a great aftertaste which makes it really easy to drink.

**醸造元**：浅間酒造株式会社
（1872年創業／群馬県）
**種類**：日本酒 特別純米酒
**度数**：15度
**精米歩合**：60%

**Brewery** : Asama sake brewing Co., Ltd. （Since 1872／Gunma）
**Kind** : Special Junmai-shu
**Alcohol Content** : 15%
**Polishing Ratio** : 60%

## 主な行き方 Major Directions

羽田空港＝（20分）⇒浜松町駅＝（5分）⇒東京駅＝（50分）⇒高崎駅＝（60分）⇒長野原草津口

Haneda Airport＝(20 min)⇒Hamamatsucho Station＝(5 min)⇒Tokyo Station＝(50 min)⇒Takasaki Station＝(60 min)⇒Naganoharakusatsuguchi

### ▶①湯畑

温泉街の中心に位置する湯畑は、草津温泉のシンボル。湯けむりが立ち上る幻想的な風景を楽しむことができるこの場所は、夕暮れ時から湯畑のライトアップが点灯され、昼と夜で異なる顔を見せます。

### ▶①Yubatake

Yubatake is the symbol of Kusatsu locating at the middle of the city. You can enjoy a mysterious view of the smoke standing up and at night the city would be lighted up changing its feature different from the daytime.

### ▶②西の河原公園

至る所から源泉が湧き出し、温泉の川がある珍しいスポット。遊歩道としても整備されていて、観光に訪れた方々の散策コースとなっています。

### ▶②Sainokawara park

There are sources of hot spring everywhere and has a lake made of hot water. Visitors can also walk along the path to explore.

**Brewery's recommended spot**
**蔵元おすすめスポット**

## 八ッ場ダム

吾妻郡長野原町川原畑
Kawarahata, Naganohara-machi, Agatsuma-gun

1967年の建設開始から52年を費やし、ようやく2019年中に完成予定のダムである。やんばツアーズとして工事現場の見学ができる（見学無料・要予約）。
出所：国土交通省関東地方整備局ホームページ（http://www.ktr.mlit.go.jp/yanba/）

## Yamba Dam

After 52 years since the dum was started to built in 1967, it is planned to be completed in 2019. You can visit here as a tour. (free to tour, need reservations)

**旅酒が見つかる場所** **Shop**

- **本多みやげ店**
  吾妻郡草津町草津110
  0279-88-2155
- **道の駅 草津運動茶屋公園特産ショップ**
  吾妻郡草津町大字草津2-1
  0279-88-0881
- **土屋商店**
  吾妻郡草津町草津386
  0279-88-2257
- **浅間酒造観光センター**
  吾妻郡長野原町長野原1392-10
  0279-82-2045
- **ホテル一井**
  吾妻郡草津町411
  0279-88-0011
- **草津温泉ホテル櫻井**
  吾妻郡草津町465-4
  0279-88-3211
- **ホテル＆スパリゾート中沢ヴィレッジ**
  吾妻郡草津町草津618
  0279-88-3232

※最新の販売店情報はホームページでご確認ください。
Please check our website for the latest retail shop information.

**◆他にも素敵な場所があります。あなただけの旅の宝を探してみましょう!**
**There are other wonderful places to visit. Let's take a trip to find a treasure of your own!**

29

FUJISAN・FUJIGOKO -Yamanashi-

# 富士山・富士五湖（山梨県）

日本を象徴する山であり、世界文化遺産ともなっている富士山。その山麓に位置する5つの湖、富士五湖。富士五湖越しに眺める富士山は、それぞれ特徴も異なり、絶景を楽しむことができます。雄大な自然を生かした様々なレジャー施設も観光客を魅了しています。

Mt. Fuji is the most famous mountain in Japan, and is also registered as the World's Heritage Site. There are five lakes at the foot of Mt. Fuji. The views of Mt. Fuji from the Fuji Five Lakes have their own features and you can enjoy superb panoramas. Various leisure facilities that make use of the magnificent nature also attract a lot of tourists.

## 旅酒 TABI-SAKE

FUJISAN・FUJIGOKO

澙富士北麓の酒米と富士山の伏流水で醸したこだわりの純米酒。柔らかな香り、きめ細やかななめらかさ、しっかりとした旨み。さまざまな食とのマリアージュを楽しむのに最適な食中酒。

Freshly brewed the Junmai-shu made with rice at Fuji North Foot and the underground water of Mt. Fuji Soft aroma, fine-grained smoothness, solid umami.

**醸造元：**井出醸造店
（1850年創業／山梨県）
**種類：**日本酒 純米酒
**度数：**15度以上16度未満
**精米歩合：**60%

**Brewery :** Ide Sake Brewery（Since 1850／Yamanashi）
**Kind :** Junmai-shu
**Alcohol Content :** 15-16%
**Polishing Ratio :** 60%

## 主な行き方 Major Directions

羽田空港＝（20分）⇒品川駅＝（20分）⇒新宿駅＝（特急60分）⇒大月駅＝（50分）⇒河口湖駅

Haneda Airport（20 min）⇒Shinagawa Station＝（20 min）⇒Shinjuku Station＝（express 60 min）⇒Otsuki Station＝（50 min）⇒Kawaguchiko Station

羽田空港＝（20分）⇒浜松町駅＝（5分）⇒東京駅／バス＝（120分）⇒河口湖駅

Haneda Airport＝（20 min）⇒Hamamatsucho Station＝（5 min）⇒Tokyo Station／Bus＝（120 min）⇒Kawaguchiko Station

## 旅の宝 Treasure Hunt

### ▶①富士山登山

複数のルートから富士山に登頂することができます。整備された登山道は登山ビギナーの方でも登りやすく、山頂からの景色は絶景の一言では表しきれない魅力があります。

### ▶②富士五湖

富士五湖は東から山中湖、河口湖、西湖、精進湖、本栖湖とあり、すべての湖が富士箱根伊豆国立公園に指定されています。それぞれの湖が個性を持ち、四季折々の景観を有しています。富士の雄姿と美しい水をたたえた湖がつくり出す絶景は時の経つのを忘れるほど感動的です。富士五湖と富士山の間には青木ヶ原樹海が広がり、手つかずの大自然も体感できます。

### ▶①Climbing Mt. Fuji

There are many routes to climb Mt. Fuji. The routes are all paved and are easy to climb for beginners. From the pinnacle of Mt. Fuji, you can see an amazing view that you can't explain in one word.

### ▶②Fuji Five Lakes (Fujigoko)

The fujigoko is five lakes: from the east, Lake Yamanaka, Lake Kawaguchi, Lake Saiko, Lake Shojiko, and Lake Motosuko. They are all registered in the Fuji-Hakone-Izu National Park. It has different sceneries in different seasons. The view of the large moutain and the beautiful water is so amazing that you would forget time. Between the mountain and the lake, there is the Aokigahara Jukai Forest where you can experience nature not being touched by humans.

**Brewery's recommended spot**
**蔵元おすすめスポット**

## 北口本宮冨士浅間神社

富士吉田市上吉田
Kamiyoshida, Fujiyoshida-shi

富士山世界文化遺産の構成資産である北口本宮冨士浅間神社の荘厳なたたずまいの社殿と樹齢千年の御神木の存在感は圧倒的。

## Kitaguchi Hongu Fuji Sengenjinja Shrine

The presence of the main shrine of the north exit Hongu Fuji Sengenjinja Shrine, a component of Mt. Fuji's World's Cultural Heritage, and the sacred tree 1,000 years old is overwhelming.

## 旅酒が見つかる場所 Shop

- **河口湖ハーブ館**
  南都留郡富士河口湖町船津6713-18
  0555-72-3082
- **西湖いやしの里 根場**
  南都留郡富士河口湖町西湖根場2710
  0555-20-4677
- **道の駅 なるさわ**
  南都留郡鳴沢村8532-63
  0555-85-3900
- **原町屋酒店**
  南都留郡山中湖村平野116
  0555-65-8015
- **道の駅 すばしり**
  静岡県駿東郡小山町須走338-44
  0550-75-6363
- **GatewayFujiyama 河口湖店**
  南都留郡富士河口湖町船津3641 河口湖駅構内
  0555-72-2214
- **忍野しのびの里**
  南都留郡忍野村忍草2845
  0555-84-1122

※最新の販売店情報はホームページでご確認ください。
Please check our website for the latest retail shop information.

◆**他にも素敵な場所があります。あなただけの旅の宝を探してみましょう!**
**There are other wonderful places to visit. Let's take a trip to find a treasure of your own!**

30

## OKINAWA ISLAND -Okinawa-

# 沖縄本島 （沖縄県）

日本最南端、青い海、澄んだ空、さんご礁……。南国的な自然の景観に恵まれた沖縄は誰しもが訪れたくなる魅力が詰まっています。独自の伝統的文化を感じる旅も良し、大自然の中で体を動かすも良し、美味しい料理や温かい地元の方との交流も良し。数々の調査で人気観光地としてトップを獲得している沖縄を味わい尽くしましょう。

The southernmost in Japan, blue sea, beautiful sky, and the coral reef.... Okinawa, where abundant tropical natures exist, is a place everyone wants to visit. Experiencing the original traditional culture, moving around in the great nature, and interaction with people and food are all attractions of Okinawa. Please enjoy Okinawa that receives the top rating as a famous tourist spot in many examination.

## 旅酒 TABI-SAKE
### OKINAWA ISLAND

10年の長い月日を重ねることで甘い熟成香が立ち上がる古酒。口に含めばコクもふくらみもたっぷりで、カカオのような優しい甘みがあります。

By spending a time of 10 years, the old Awamori has a sweet scent. When you have it in your mouth, it has a deep taste and has a sweetness like cacao.

**醸造元**：有限会社山川酒造
（1946年創業／沖縄県）
**種類**：琉球泡盛 10年古酒
**度数**：30度

**Brewery** : Yamakawa Shuzo Co., Ltd.
（Since 1946／Okinawa）
**Kind** : Ryukyu Awamori 10 years old
**Alcohol Content** : 30%

## 主な行き方 Major Directions

羽田空港＝（165分）⇒那覇空港
Haneda Airport＝（165 min）⇒Naha Airport

## 旅の宝 Treasure Hunt

### ①那覇

沖縄の中心地である那覇。琉球王国の中心である首里城や、観光客でにぎわう国際通り、もちろん沖縄を象徴するような綺麗な海にも入ることが出来ます。

### ①Naha

Located in the middle of Okinawa. People can visit the Shuri Castle which was the heart of the Ryukyu Kingdom, the Kokusai Street where it is filled with visitors, and of course the beautiful ocean that represents Okinawa.

### ②恩納村

沖縄のリゾート地として有名な恩納村は、那覇市の北に位置しています。サンゴ礁が生息する綺麗な海が見所となっており、ビーチや琉球村などの歴史文化を楽しむことが出来ます。

### ②Onnason Village

Onnason, famous for its resort area, is located in the north of Naha. The beautiful ocean where coral reef exists is one of the highlights, and people can enjoy the historical culture of Okinawa by visiting the beach and the Ryukyu Village.

## Brewery's recommended spot
## 蔵元おすすめスポット

### 海洋博公園

国頭郡本部町石川424

424, Ishikawa, Motobu-cho Kunigami-gun

沖縄美ら海水族館、エメラルドビーチなど、沖縄のみどころが集まる公園。

### Ocean Expo Park

It is a park where it has many attractions of Okinawa such as the Churaumi-Aquarium and the Emerald Beach.

### 海洋博公園サマーフェスティバル

国頭郡本部町石川424

424, Ishikawa, Motobu-cho Kunigami-gun

### Ocean Expo Park Summer Festival

A large event held in the Emerald beach.

エメラルドビーチにて開催される県内最大級のイベント。

## 旅酒が見つかる場所　Shop

- **おきなわ屋本店**
  那覇市牧志1-2-31
  098-860-7848
  ※上記本店をはじめ、おきなわ屋各店で取扱い。
- **泡盛館**
  那覇市首里寒川町1-81
  098-885-5681
- **道の駅 許田**
  名護市許田17-1
  0980-54-0880
- **キラク花城**
  那覇市牧志3-5-4
  098-863-1110
- **田空の駅 ハーソー公園**
  国頭郡本部町具志堅1334
  0980-48-3835
- **那覇空港国内線ターミナルビル さくら売店**
  那覇市鏡水150
  098-840-1473
- **ホテル オリオン モトブ リゾート＆スパ**
  国頭郡本部町備瀬148-1
  0980-51-7300
- **道の駅 ゆいゆい国頭 国頭村観光物産センター**
  国頭郡国頭村字奥間1605
  0980-41-5555
- **喜屋武商店**
  那覇市前島3-4-16
  098-868-5270
- **わしたショップ国際通り本店**
  那覇市久茂地3-2-22
  JAドリーム館
  098-864-0555
- **ショップやまかわ**
  本部町並里59
  0980-47-5651
- **琉宮城蝶々園**
  本部町山川390-1
  0980-47-3456
- **羽地の駅**
  名護市真喜屋763
  0980-58-2358
- **JAファーマーズマーケットやんばる**
  名護市宮里4-6-37
  0980-50-9885

※最新の販売店情報はホームページでご確認ください。
Please check our website for the latest retail shop information.

（新潟県）

31

日本有数の米どころとして有名な越後。越後の戦国武将・上杉謙信の居城があった上越市、妙高山を望む高原リゾートに複数の温泉街がある妙高市、スキー場の玄関口として有名な湯沢町など、上越妙高駅・越後湯沢駅を中心に、美しい越後の国をお楽しみください。

Echigo is famous as one of the best rice producing areas in Japan. Joetsu-shi is where Uesugi Kenshin, a warrior in Echigo, had his castle in the age of provincial wars. Myoko-shi is a highland resort overlooking Mt. Myoko and has several spa resorts. Yuzawa-cho is famous as a gateway to a ski resort. Each place is accessible from either Joetsu-myoko Station or Echigo-yuzawa Station. You can enjoy the area of Echigo at your leisure, as it offers sightseeing activities and beautiful nature.

## 旅 TABI-SAKE 酒

### ECHIGO

上品な麹の香りと旨み豊かな味わい。スッキリとした口あたりで冷やして良し、燗して良し。幅広い料理と引き立て合う気品のある淡麗辛旨口の特別純米酒。

An elegant scent of malter rice and a rich flavor. It has a very refreshing taste and goes well with both cold and hot. The dry bitter taste of the sake goes along well with all types of food.

**醸造元**：妙高酒造株式会社
（1815年創業／新潟県）
**種類**：日本酒 特別純米酒
**度数**：15度
**精米歩合**：60%

**Brewery** : Myoko Shuzo Co., Ltd.
(Since 1815／Niigata)
**Kind** : Special Junmai-shu
**Alcohol Content** : 15%
**Polishing Ratio** : 60%

甘辛度
Sweet-dryness

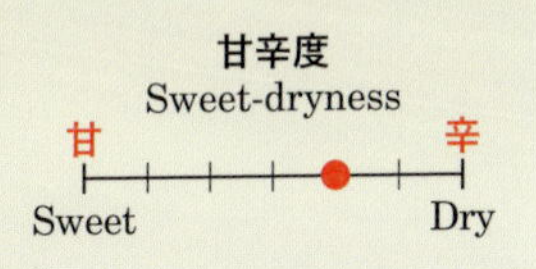

## 主な行き方 Major Directions

羽田空港＝(20分)⇒浜松町＝(5分)⇒東京駅＝(新幹線115分)⇒上越妙高駅

Haneda Airport＝(20 min)⇒Hamamatsucho Station＝(5 min)⇒Tokyo Station＝(Shinkansen 150 min)⇒Joetsu Myoko Station

羽田空港＝(20分)⇒浜松町＝(5分)⇒東京駅＝(新幹線80分)⇒越後湯沢駅

Haneda Airport＝(20 min)⇒Hamamatsucho Station＝(5 min)⇒Tokyo Station＝(Shinkansen 80 min)⇒Echigo-Yuzawa Station＝(75 min)

## 旅の宝 Treasure Hunt

### ▶①春日山城

戦国武将上杉謙信の城として知られる春日山城。複雑な自然の地形を利用した堅固な城塞から、難攻不落な天下の名城といわれていたこの城跡は、日本百名城にも数えられる国指定遺跡です。

### ▶①Kasugayama Castle

Kasugayama Castle is known for the warriors, Kenshin Uesugi's castle. Using the complicated land, it is very defensive and is famous for enemies impossible to break in. The castle is one of the famous 100 castles in Japan and is registered as the country's historical spots.

### ▶②高田公園

日本三大夜桜で知られる高田城跡の都市公園。上越市の中心に位置し、約4000本の桜がお堀に沿ってライトアップされる時期には、特に多くの観光客が訪れます。

### ▶②Takada Park

The park is famous for the night cherry blossoms. It is located at the middle of the city and almost 4000 cherry blossom trees are planted and many visitors come during the lightening up seasons.

**Brewery's recommended spot**
**蔵元おすすめスポット**

## 越後・謙信SAKEまつり

上越市本町
Honcho, Joetsu-shi

毎年10月に行われる。上越地域の日本酒をはじめ、ワイン、どぶろく、地ビールなど「SAKE」を一堂に集め、上越地域の酒づくり文化を紹介するイベント。

## Echigo Kenshin Sake Festival

It is held on October every year. Starting with the Japanese sake from the Joetsu region, there are many other types such as wine, doburoku, and beer. It is an event to introduce the culture of sake making in the Joetsu region.

## 旅酒が見つかる場所 Shop

- **妙高酒造 小売り店部**
  上越市南本町2-7-47
  025-522-2111
- **さくら百嘉店 上越妙高駅店**
  上越市大和2-1-1
  025-520-9758
- **カンパーナあらい かんずり工房 越後の地酒**
  妙高市大字猪野山58-1
  0255-70-1882
- **カネタみやげ店**
  妙高市田口315
  0255-86-2307
- **道の駅 新潟ふるさと村**
  バザール館1F 新潟の酒コーナー
  新潟市西区山田2307
  025-230-3000

※最新の販売店情報はホームページでご確認ください。
Please check our website for the latest retail shop information.

**◆他にも素敵な場所があります。あなただけの旅の宝を探してみましょう!**
**There are other wonderful places to visit. Let's take a trip to find a treasure of your own!**

32

# HAKONE -Kanagawa-

# 箱根（神奈川県）

神奈川県足柄下郡に位置し、首都圏からほど近い温泉地・観光地として知られている箱根。関東からはもちろん、海外からも毎年多くの観光客が訪れています。周辺には、自然の雄大さを感じる景勝地や、美術館、博物館、寺社仏閣などもあり、何日間滞在しても飽きることがありません。

Hakone is located in Kanagawa prefecture near Ashigara shimo district. It is famous for its spa resort close to the metropolis. Not only people from the Kanto region but also people from outside of Japan visit here. In proximity of the town, there are the great nature, museums and galleries, and several temples and shrines so tourists would not lose interest even though they stay here for many days.

**HAKONE**

柑橘系の爽やかで軽快な香りと淡麗な味わいが調和した純米吟醸酒。

The Junmai Ginjo-shu having a harmony of the refreshing citrus fragrance and the crispy taste.

**醸造元**：井上酒造株式会社
（1789年創業／神奈川県）
**種類**：日本酒 純米吟醸酒
**度数**：15度
**精米歩合**：55%

**Brewery** : Inoue Shuzo Co., Ltd.
(Since 1789／Kanagawa)
**Kind** : Junmai Ginjo-shu
**Alcohol Content** : 15%
**Polishing Ratio** : 55%

**甘辛度**
Sweet-dryness

甘 Sweet ―――――●―― 辛 Dry

主な行き方 Major Directions

羽田空港＝(20分)⇒品川駅＝(30分)⇒小田原駅＝(15分)⇒箱根湯本駅

Haneda Airport(20 min)⇒Shinagawa Station＝(30 min)⇒Odawara Station＝(15 min)⇒Hakone Yumoto Station

旅の宝 Treasure Hunt

### ▶①大涌谷

至る所から白い煙が立ち上る景観で、江戸時代までは地獄谷とも呼ばれた大涌谷。ロープウェイに乗って絶景をみたら、黒く変色した「黒たまご」や温泉を楽しみましょう。

### ▶①Owakudani

White smokes are rising up in the air from many places, and until the Edo period Owakudani was called as the Jigokudani (hell valley). After looking down the view from the ropeway, tourists should check the “black eggs” and the hot spring.

### ▶②芦ノ湖

県内で最大の湖である芦ノ湖。湖畔からは富士山を望むことができ、周辺にはホテルや温泉、観光施設などが建ち並ぶリゾートエリアです。自然や歴史を感じながら、優雅な時間を過ごすことができるでしょう。観光名所・パワースポットとして有名な箱根神社にも行くことができます。

### ▶②Lake Ashinoko

Lake Ashinoko is the largest lake in Kanagawa prefecture. From the lakeside, tourists can see the view of Mt. Fuji. Around the lake, there are many hotels, spas, and tourist resorts. Tourists could spend a great time feeling the nature and the history. They can also go to the Hakonejinja Shrine which is famous for a place where people can feel mysterious powers.

### Brewery's recommended spot 蔵元おすすめスポット

## 旧街道石畳

足柄下郡箱根町畑宿
Hatajuku, Hakone-machi, Ashigarashimo-gun

江戸時代に整備された旧東海道に残る石畳。通称「箱根八里」で知られる天下の難所だった。
現在も当時を偲ぶ風景が残り、江戸時代の雰囲気を感じられるスポット。

## Stone-paved Road

It is a stone pavement left from the Edo period. It was a dangerous spot known as "Hakone-Hachiri." It still has its features from the time and you can enjoy the Edo atmosphere.

### 旅酒が見つかる場所 Shop

- **酒岳堂**
  足柄下郡箱根町湯本637
  0460-85-5328
- **ローソン 箱根仙石高原店**
  足柄下郡箱根町仙石原1246-265
  0460-86-3941
- **ローソン 箱根二ノ平店**
  足柄下郡箱根町二ノ平1202-14
  0460-83-8898
- **ローソン 箱根宮ノ下店**
  足柄下郡箱根町宮ノ下376
  0460-87-7611

※最新の販売店情報はホームページでご確認ください。
Please check our website for the latest retail shop information.

**◆他にも素敵な場所があります。あなただけの旅の宝を探してみましょう!**
**There are other wonderful places to visit. Let's take a trip to find a treasure of your own!**

## KISHU・KOYASAN -Wakayama-

# 33 紀州・高野山（和歌山県）

高野山をはじめ、自然豊かな観光名所が多くある和歌山県。高野山は、世界遺産にも登録されている日本仏教の聖地です。金剛峯寺をはじめとして数多くの寺院があり、高野山全体がお寺であるという見方をされています。
1年中温暖な気候のリゾート地、南紀白浜も観光地として有名で、温泉や海水浴などを楽しむことができ、別荘地としても知られています。

Wakayama prefecture has many scenic spots rich in nature including Mt. Koya. Mt. Koya is a sacred place of Japanese Buddhism and has been designated as a World Heritage Site. There are many temples including Kongobuji Temple, and the whole mountain is considered to be one huge temple. Nanki Shirahama, a resort area with a warm climate throughout the year, is also a famous tourist spot, where you can enjoy hot springs and sea bathing. It is also known as an area of holiday villas.

## 旅 TABI-SAKE 酒

### KISHU ・ KOYASAN

紀州南高梅を使用して丹念に漬け込んだ本格梅酒。南高梅ならではの、甘みも酸味も非常に濃い味わいです。ロックやソーダ割りでも梅の味わいを存分に感じていただける、昔ながらの梅酒です。

It is a genuine umeshu that uses the Kishunanko-plums. The sweetness and sourness are both rich. Drinking it with just ice or with soda makes you feel the flavor of the plum fully.

**醸造元**：中野BC株式会社
（1932年創業／和歌山県）
**種類**：梅酒
**度数**：14度

**Brewery** : Nakano BC Co., Ltd.
（Since 1932／Wakayama）
**Kind** : Plum Wine
**Alcohol Content** : 14%

## 主な行き方 Major Directions

羽田空港＝(65分)⇒大阪国際空港／バス＝(30分)⇒南海難波駅＝(80分)⇒極楽橋駅＝(5分)⇒高野山駅

Haneda Airport＝(65 min)⇒Osaka International Airport／Bus＝(30 min)⇒Nankai Namba Station＝(80 min)⇒Gokurakubashi Station＝(5 min)⇒Koyasan Station

羽田空港＝(75分)⇒南紀白浜空港／バス＝(30分)⇒白浜駅

Haneda Airport＝(75 min)⇒Nanki Shirahama Airport／Bus＝(70 min)⇒Shirahama Station

## 旅の宝 Treasure Hunt

### ▶①熊野本宮大社

熊野本宮大社は、熊野三山と呼ばれる3つの神社の中心であり、今もなお多くの人々の信仰を集めています。「紀伊山地の霊場と参詣道」として世界遺産に登録されており、建造物等の一部は重要文化財にもなっています。厳かで、悠久の歴史を感じることのできる場所です。

### ▶②瀞峡

瀞峡は、四季折々の景色を堪能することができる大峡谷です。ウォータージェット船に乗って瀞峡へ行けば、コバルトブルーに澄み渡る川や、青々とした木々、紅葉、そしてダイナミックな断崖をみることができます。国の名勝に指定されている景色をぜひご覧ください。

### ▶①Kumanohongutaisha Shrine

Kumanohongutaisha Shrine is one of the three famous shrines in Kumanosanzan and many people still visit here. It is also registered as the World's Heritage and one of the building is the country's Important Cultural Property. It is a very austere place where people can experience the history.

### ▶②Dorokyo Gorge

Dorokyo is a large valley where tourists can enjoy different views for each season. If tourists get on a boat and go to the Dorokyo Gorge, they can glance the clear blue river, green trees, and red leaves. Please enjoy the view of a dynamic cliff registered as the country's beauty spot.

**Brewery's recommended spot**
**蔵元おすすめスポット**

## 丹生川の鯉のぼり

伊都郡九度山町入郷5-5
5-5, Nyugo, Kudoyama-cho Ito-gun

丹生川と紀の川の合流する丹生橋付近で、毎週開催される「こいのぼりの丹生川渡し」。
約100匹以上のこいのぼりが空を泳ぐ。

## The Koinobori of Nyukawa River

The Koinobori festival is held every week near the Nyu Bridge which is where the rivers merge. About 100 carps are like swimming in the sky.

写真提供：公益社団法人和歌山観光連盟

## 円月島の夕日

西牟婁郡白浜町3740
3740, Shirahama-cho, Nishimuro-gun

白浜にある島の中央に円月形の海蝕洞が空いている特徴的な島。
円月島に沈む夕陽は「日本の夕陽100選」に選ばれている。

## The Sunset of Engetsu Island

It is an island that has a moon shaped whole in the middle of the island. The sunset of Engetsu Island is selected as the 100 famous sunsets of Japan.

**旅酒が見つかる場所** **Shop**

- **勝間屋**
  伊都郡高野町高野山782
  0736-56-2334
- **とれとれ市場**
  西牟婁郡白浜町堅田2521
  0739-42-1010

※最新の販売店情報はホームページでご確認ください。
Please check our website for the latest retail shop information.

34

# 知多 （愛知県）

伊勢湾に面し、東は知多湾・三河湾に挟まれた知多半島。三方を海に囲われたこのエリアは、海の魅力をふんだんに活かした海産物が魅力的。温暖な気候で、マリンスポーツなどの自然を生かしたスポーツも楽しめます。自動車産業をはじめとするものづくりの地域としても有名な地域で、有名な焼き物も魅力のひとつです。

Facing Ise Bay, Chita Bay and Mikawa Bay, the Chita Peninsula is surrounded by the sea on three sides. It is popular for its seafood that brings out the charm of the sea. The mild climate allows you to enjoy sports with nature such as marine sports. This area is also famous as a manufacturing region including the automobile industry. Pottery is also thriving here.

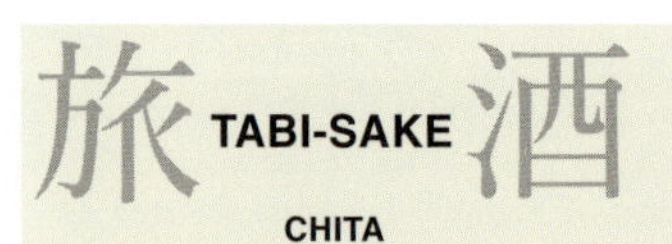


お米由来の風味、旨みと、スッキリ辛口な味わいが特徴です。

It features flavor and umami derived from rice and a refreshing taste.

**醸造元**：中埜酒造株式会社
（1844年創業／愛知県）
**種類**：日本酒 純米酒
**度数**：15度
**精米歩合**：65 %

**Brewery** : Nakano Sake Brewery Co., Ltd.（Since 1844／Aichi）
**Kind** : Junmai-shu
**Alcohol Content** : 15%
**Polishing Ratio** : 65%

甘辛度
Sweet-dryness
甘 ―――――●―― 辛
Sweet　　　　　Dry

## 主な行き方 Major Directions

羽田空港＝(20分)⇒品川駅＝(95分)⇒名古屋駅＝(40分)⇒知多半田駅

Haneda Airport＝(20 min)⇒Shinagawa Station＝(95 min)⇒Nagoya Station＝(40 min)⇒Chitahanda Station

### ①恋の水神社

縁結びのパワースポットとして人気の恋の水神社。多くの若者で賑わっています。

### ①Koinomizujinja Shrine

The shrine is famous for having a mystical power of matchmaking. Many young people visit here.

### ②やきもの散歩道

焼き物の町、常滑の文化を歩きながら味わえるやきもの散歩道。さまざまなモニュメントが連なり、退屈させない散歩道となっています。

### ②Pottery Footpath

The Pottery Footpath is a path where you can enjoy the Tokoname culture while walking. There are many monuments by the street and you won't get bored just walking.

**Brewery's recommended spot**
**蔵元おすすめスポット**

## 日間賀島

知多郡南知多町日間賀島
Himakajima, Minamichita-cho, Chita-gun

三河湾に浮かぶ離島。イルカとのふれあいや漁業体験等、海遊びが楽しめる島。タコとフグが有名だが、その他の海の幸も豊富。

## Himakajima Island

It is an island floating on the Mikawa Bay. You can play with the dolphins and experience the work of fishers and have great time at the ocean. It is famous for octopus and puffer fish but other products of the sea are famous.

## 旅酒が見つかる場所 Shop

- **酒の文化館**
  半田市東本町2-24
  0569-23-1499
- **かもめ売店**
  知多郡南知多町日間賀島西浜5
  0569-68-2870
- **日間賀観光ホテル**
  知多郡南知多町日間賀島
  0569-68-2211

※最新の販売店情報はホームページでご確認ください。
Please check our website for the latest retail shop information.

◆**他にも素敵な場所があります。あなただけの旅の宝を探してみましょう!**
**There are other wonderful places to visit. Let's take a trip to find a treasure of your own!**

35

# KURASHIKI -Okayama-

# 倉敷（岡山県）

岡山県南部に位置する倉敷。徳川幕府の直轄地「天領」として栄え、今でもその時代の風情を残した観光名所です。白壁の町、文化の町として有名な倉敷には、美観地区の古き情緒を今も受け継ぐ町並みだけではなく、「ひやさい」と呼ばれる路地裏にも数多くの隠れた名店があり、様々な魅力を持つ町です。

Kurashiki is located at the south of Okayama prefceture. It still has some features of the region under the control of the Tokugawa. Kurashiki, known as a town of culture, is famous for its aesthetic area that retains a traditional townscape with white-walled warehouses. The town has various fascinating features and once you get in the back alleys, which are called "Hiyasai," you can find quiet shops and restaurants with a cozy ambience.

## 旅 TABI-SAKE 酒

### KURASHIKI

雄町米は現存する日本最古の酒造好適米で、地元岡山で95%生産されています。その雄町米を100%使用した純米酒です。丸みのあるボディと旨みが交わるまろやかなお酒です。

Omachi rice is the oldest type of rice and Okayama produces 95% of it by itself. It uses 100% of the Omachi rice. It is the mild Junmai-shu which has a round body and rich umami.

**醸造元**：熊屋酒造有限会社
（1716年創業／岡山県）
**種類**：日本酒 特別純米酒
**度数**：15%
**精米歩合**：60%

**Brewery** : Kumaya Shuzo Co., Ltd.
(Since 1716／Okayama)
**Kind** : Special Junmai-shu
**Alcohol Content** : 15%
**Polishing Ratio** : 60%

**甘辛度**
Sweet-dryness

甘 Sweet ——●— Dry 辛

## 主な行き方 Major Directions

羽田空港＝（75分）⇒岡山桃太郎空港／バス＝（35分）⇒倉敷駅
Haneda Airport＝(75 min) ⇒Momotaro Okayama Airport／Bus ＝(35 min) ⇒Kurashiki Station

## 旅の宝 Treasure Hunt

### ▶①大原美術館

日本で初めての私立西洋近代美術館。海外の名美術品が多数集まる、日本を代表する美術館のひとつです。

### ▶②倉敷美観地区

国の重要伝統的建造物群保存地区として指定されている名所です。街道一帯に白壁なまこ壁の屋敷や蔵が並び、天領時代の町並みが今もなお残っています

### ▶①Ohara Museum of Art

It was the first private western museum of modern art in Japan. It is one of the famous museum in Japan where many arts are collected from the world.

### ▶②Kurashiki Bikan Historical Quarter

It is an important preservation districts for groups of historic buildings for Japan. The whole street is designed with storehouses white walls covered with square tiles and still has features remaining from the Tenryo period.

Brewery's recommended spot
蔵元おすすめスポット

## 日本第一熊野神社

倉敷市林684
684, Hayashi, Kurashiki-shi

全国に3800あまりある熊野神社のなかでも、本宮（紀州熊野本宮大社）の昔の姿を残している。
「いじめ除け」の神としての八尾羅宮が境内にあり、広く信仰を集めている。

## Nihon Daiichi Kumanojinja Shrine

There are about 3800 shrines of the Kumano, but this one is the main Kumano Shrine. It is the god of no bullying and Happiragu is inside the area.

ツタ（＝アイビー）のからまる赤いレンガが目を引く、明治時代の倉敷紡績所発祥工場を再利用して生まれた複合文化施設。敷地内には、陶芸が体験できる工房やホテルがあり、時代を超えて受け継がれる近代産業遺産である。

## 倉敷アイビースクエア

倉敷市本町7-2
7-2, Hon-machi, Kurashiki-shi

## Kurashiki Ivy Square

It is a complex cultural facility created by renovating the original factory of Kurashiki Spinning in the Meiji period, which features ivy-covered red bricks. There are a hotel and a workshop where you can experience ceramic art on the premises. It is a modern industrial heritage that has been passed down through the ages.

## 旅酒が見つかる場所 Shop

**天満屋 倉敷店**
倉敷市阿知1-7-1
086-426-2796

**おみやげ街道 倉敷**
倉敷市阿知1-1-1
JR倉敷駅構内
086-424-6606

※最新の販売店情報はホームページでご確認ください。
Please check our website for the latest retail shop information.

36

# MITO・OARAI -Ibaraki-

# 水戸・大洗 （茨城県）

水戸黄門でお馴染みの徳川光圀公で有名な水戸。そして水戸から東に進むと関東有数の海水浴場と豊富な漁場を持つ大洗。茨城県の観光名所として名高いこのエリアには、素晴らしい景観とさまざまな美味しい魚介料理を楽しむことができ、多くの観光客が訪れています

Mito is famous for Mitsukuni Tokugawa, who is popular in Mito Komon. When you move a little to the east, there is Oarai, where there is famous beaches and many sea ports in the Kanto region. Many tourists visit this area where you can enjoy the beautiful sceneries and the seafood.

## 旅 TABI-SAKE 酒

MITO ・ OARAI

山田錦の親米にあたる短稈渡船。複雑な米の旨みと、ほのかな酸味が特徴です。全量短稈渡船を使い醸した、純米吟醸酒です。芳醇な吟醸香、口に入れたときのふくらみをお楽しみください。

Tankan wataribune is mother cultivars of Yamadanishiki, which is the most cammonly grown sake-rice. It has a complicated taste of rice and a little bit of sourness. It is the Junmai Ginjo-shu using the Tankan Wataribune. Please enjoy the spread of the flavor in your mouth.

**醸造元：** 吉久保酒造株式会社（1790年創業／茨城県）
**種類：** 日本酒 純米吟醸酒
**度数：** 16%
**精米歩合：** 50%

**Brewery** : Yoshikubo Brewery Co., Ltd.（Since 1790／Ibaraki）
**Kind** : Junmai Ginjo-shu
**Alcohol Content** : 16%
**Polishing Ratio** : 50%

甘辛度
Sweet-dryness

## 主な行き方 Major Directions

羽田空港＝(20分)⇒品川駅＝(80分)⇒水戸駅＝(15分)⇒大洗駅

Haneda Airport＝(20 min)⇒Shinagawa Station＝(80 min)⇒Mito Station＝(15 min)⇒Oarai Station

## 旅の宝 Treasure Hunt

### ▶①偕楽園

岡山市の後楽園や金沢市の兼六園と並んで、日本三名園のひとつに数えられてきた日本庭園。自然を生かしたその作りは、日本の美と文化を味わうことができます。

### ▶①Kairakuen Garden

Kourakuen in Okayama, Kenrokuen in Kanazawa, and Kairakuen in Mito are the three famous Japanese gardens. The garden uses the natural feature itself so take a look at the Japanese beauty and culture.

### ▶②大洗海岸

茨城エリア最大のビーチ、大洗海岸。海水浴以外の夏の遊びや、新鮮な海鮮グルメも楽しむことができます。

### ▶②Oarai Beach

Oarai beach is the largest beach in the Ibaraki area. You can enjoy not only swimming in the beach but also other summer activities and seafood gourmet.

Brewery's recommended spot
蔵元おすすめスポット

## 水戸黄門まつり

水戸市中心部と千波湖周辺
around the main area of Mito-shi and Lake Senbako

1961年に始まった祭りで、毎年8月の第1金・土・日の3日間、開催されている。祭り中には、千波湖にて5年連続日本一に輝いた野村花火が約5000発打ち上げられ、山車巡行、神輿連合渡御、水戸黄門パレード、市民カーニバルin MITOなどのさまざまなイベントが開催される。

## Mito Komon Festival

It started in 1961, and is held on the first Friday, Saturday, and Sunday of August every year. During the festival, at Lake Senbako, about 5000 Nomura fireworks, which is the best fireworks in Japan for 5 consecutive years, are shot, festival car cruise, Omikoshi, Mito Komon parade, and the carnival in MITO are held.

## 吉田神社 秋季例大祭

水戸市宮内町3193-2
3193-2, Miyauchi-cho, Mito-shi

吉田神社は、日本武尊（ヤマトタケルノミコト）を祀る神社であり、日本武尊が東征の際、この地で休まれたことにより創建されたといわれている。秋季例大祭は古くから続き、毎年10月中旬に神輿渡御が盛大に行われる。また、神社境内で夜祭が行われ、神輿7台が集結し、お囃子が競演をする。笛や太鼓のお囃子や狐の舞による山車が華やかさを彩る。

## Annual Festival of Yoshida-jinja Shrine

The Yoshidajinja Shrine is where Yamatotakeru-no-mikoto is presented, and when he visited this area he rested at this shrine. The Autumn Festival is held from a long time ago. It is held on Sat and Sun every year near the 15 and 16th of October every year and is a big event. At Friday, there are night festivals in the area, which 7 mikoshis gather and feature together. The whistles and taiko are played and is flamboyant.

### 旅酒が見つかる場所 Shop

- **おみやげやプラム水戸**
  水戸市宮町1-1-1
  029-228-9363
- **Petit LOCO'S（プチ ロコス）**
  水戸市宮町1-7-31
  029-225-5537
- **大洗イエローポート**
  水戸市大串町897
  029-269-3009
- **大洗まいわい市場**
  東茨城郡大洗町港中央9-2
  029-266-1147
- **ジュピター水戸店**
  水戸市宮町1-1-1 2階
  029-233-3911
- **偕楽園レストハウス**
  水戸市常磐町1-3
  （常盤神社境内・偕楽園車門前）

※最新の販売店情報はホームページでご確認ください。
Please check our website for the latest retail shop information.

## ECHIZEN -Fukui-

# 越前

（福井県）

37

北陸を代表する観光名所のひとつ、越前。国の伝統的工芸品に指定されている「越前焼」「越前和紙」「越前打刃物」「越前箪笥」や、郷土料理「越前おろしそば」など、人の手で創られるさまざまな魅力にあふれており、毎年多くの観光客が訪れています。海と山に囲まれており、山海の幸が多くあり、グルメも魅力的です。

Echizen is a famous tourist spot in the Hokuriku region. It is full of charm with its own specialty handicrafts and food. “Echizen Yaki (pottery),” “Echizen Washi (Japanese paper),” “Echizen Uchi Hamono (cutlery),” and “Echizen Tansu (drawers)” are designated as national traditional crafts. “Echizen Oroshi Soba (buckwheat noodles with grated radish)” is popular as a local dish. Surrounded by the sea and mountains, this area is popular for its gourmet food with ingredients from the mountains and the sea and attracts many tourists every year.

## 旅 TABI-SAKE 酒

### ECHIZEN

爽やかな香りと越の雫（福井県酒造元のみ使用）の旨みが味わえるお酒。福井で新しく開発された酵母FK-8010の味と香りを楽しんでください。

You can enjoy the refreshing scent and the umami of Koshi no Shizuku. Please enjoy the new yeast, FK-8010, developed in Fukui prefecture.

**醸造元：**伊藤酒造合資会社

（1894年創業／福井県）

**種類：**日本酒　純米大吟醸酒

**度数：**16%

**精米歩合：**50%

**Brewery :** Ito Shuzo Limited Company

（Since 1894／Fukui）

**Kind :** Junmai Daiginjo-shu

**Alcohol Content :** 16%

**Polishing Ratio :** 50%

**甘辛度**

Sweet-dryness

甘 ―――――――――― 辛

Sweet ―――――●―― Dry

## 主な行き方 Major Directions

羽田空港＝(60分)⇒小松空港／バス＝(55分)⇒福井駅
Haneda Airport＝(60 min)⇒Komatsu Airport／Bus＝(55 min)⇒Fukui Station

## 旅の宝 Treasure Hunt

### ▶①越前がに

全国のズワイガニのなかでもトップブランドといわれる越前がに。繊細で甘みがある身がぎっしりと詰まり、甲羅のなかには、濃厚なコクがあるカニ味噌がたっぷりと入っています。

### ▶②越前焼

日本六古窯のひとつに数えられ、平安時代末期から続く古い歴史を持つ伝統工芸「越前焼」。その特徴は、越前の土の特徴を生かした素朴な頑丈さです。

### ▶①Echizen Crab

Echizen Crab is said to be one of the best crabs in japan. There are a lot of soft and sweet meat in the legs, and inside the shell, there are rich and thick crab innards.

### ▶②Echizen-yaki (Echizen Ware)

The echizen-yaki is a traditional handcraft made from the Heian period and is one of Nihon Rokkoyo (the six oid kilns of Japan). It uses the soil of Echizen, and is a strong and simple pottery.

**Brewery's recommended spot 蔵元おすすめスポット**

小さな居酒屋がおススメ

海の幸も山の幸も豊富な福井ならではの味を堪能できる場所。

Small Izakayas

A place where you can enjoy the products from the ocean and land.

## 福井県立恐竜博物館

勝山市村岡町寺尾51-11
51-11, Terao, Murokocho, Katsuyama-shi

カナダのロイヤル・ティレル古生物学博物館、中国の自貢恐竜博物館と並び、世界三大恐竜博物館と称される。
忠実に再現されたジオラマやダイナミックな恐竜骨格等迫力のある展示が多数あり、世界から注目される福井生まれの恐竜たちの展示も必見。

## Fukui Prefectural Dinosaur Museum

It is one of the three dinosaur museums in the world along with the Zigong Dinosaur Museum (China) and Royal Tyrrell Museum of Paleontology( Canada). There are many georamas and dynamic exhibitions, and dinosaurs born in Fukui prefecture are given attention.

---

1244年に道元禅師によって開創された「日本曹洞宗」の出家参禅の道場。
現在も当時のまま、修行僧が修行生活に励んでいる。厳かな禅の世界を感じることのできる場所。

## 永平寺

吉田郡永平寺町志比5-15
5-15, Shihi, Eiheiji-cho Yoshida-gun

## Eiheiji Temple

Eiheiji Temple was founded in 1244 by Dogen Zenji, a Japanese Buddhist priest, as a training monastery of the Soto school. Still today, the ascetic monks devote themselves to the practice in a manner taken over from the old days. It is a place where you can feel the solemn world of Zen.

### 旅酒が見つかる場所 Shop

- **越前酒乃店はやし 本店**
  越前市平和町12-13
  0778-22-1281
- **越前酒乃店はやし SAKE Atelie 福井春山店**
  福井市春山1-1-18
  0776-63-5120

※最新の販売店情報はホームページでご確認ください。
Please check our website for the latest retail shop information.

38

## IZU -Shizuoka-

# 伊豆 （静岡県）

日本有数の温泉街を持つ伊豆。富士山を望める美しい海、自然の中で爽やかな風を感じられる伊豆高原、河津桜等の四季折々の花々、各所に散らばるテーマパークなど、観光客を魅了するスポットが多くあります。東京からのアクセスの良さから週末旅行に訪れる人も多く、日本屈指の観光地としても知られています。

As one of the major spa resorts in Japan, Izu has many tourist attractions such as the beautiful sea with Mt. Fuji in the background, the Izu Kogen Highlands where you can feel a refreshing breeze in nature, seasonal flowers such as Kawazu-zakura, and theme parks everywhere. It is known as one of Japan's top tourist destinations with many people visiting on weekends due to its good access from Tokyo.

## 旅 TABI-SAKE 酒

### IZU

酒米"五百万石"の特性を生かし、ドライでスッキリした味わい、キレのよさ、ほのかな含み香が特長。冷やでも燗でもおいしくいただけます。カツオのたたきからウナギの蒲焼までお料理も選びません。

Using the feature of "Gohyakumangoku," it has a dry and refreshing taste with a scent spreading in the mouth. It goes well with cold or hot. It also goes along with any kind of food.

**醸造元**：三和酒造株式会社
（1686創業／静岡県）
**種類**：日本酒 純米吟醸酒
**度数**：15度以上16度未満
**精米歩合**：60%

**Brewery** : Sanwa Brewed Co., Ltd.
(Since 1686／Shizuoka)
**Kind** : Junmai Ginjo-shu
**Alcohol Content** : 15-16%
**Polishing Ratio** : 60%

**甘辛度**
Sweet-dryness

甘 Sweet ——— 辛 Dry

## 主な行き方 Major Directions

羽田空港＝(20 min)⇒品川駅＝(40分)⇒熱海駅＝(40分)⇒伊豆高原駅

Haneda Airport＝(20 min)⇒Shinagawa Station＝(40 min)⇒Atami Station＝(40 min)⇒Izukogen Station

## 旅の宝 Treasure Hunt

### ▶①城ヶ崎海岸

大室山が約4000年前に噴火したとき溶岩が海に流れ出し、海の侵食作用で削られてできた雄大な出入りの激しい溶岩岩石海岸。絶壁にかけられた吊り橋が人気を集めています。

### ▶①Jogasaki Coast

About 400 years ago the Mt. Omuro erupted and the lava flowed in the ocean. The coast is made by the ocean eroding the lava stones. The bridge hanging one the cliff is a famous touring spot.

### ▶②修善寺温泉

伊豆半島で最も歴史がある温泉、修善寺温泉。伊豆の小京都とも呼ばれ、街中を散策すれば、古都の風情に浸ることができます。

### ▶②Shuzenji Onsen

It is the oldest hot spring in the Izu region. Shuzenji Onsen is called as little Kyoto, and once you walk in the city, you can feel some traces of the historic city.

Brewery's recommended spot
蔵元おすすめスポット

## 清見寺

静岡市清水区興津清見寺町418-1
418-1, Okitsuseikenji-cho, Shimizu-ku Shizuoka-shi

臥龍梅の銘柄のいわれとなった徳川家康お手植えの臥龍梅がある。江戸時代、朝鮮通信使一向に宿を提供し、数々の記念の品を蔵することでも知られている。

## Senkenji Temple

The garyubai, which Ieyasu Tokugawa planted, are found in this area. In the Edo period, it let the Korean goodwill mission to the shogunate rest here and possessed many tokens.

## 伊豆海洋公園

伊東市富戸841-1
841-1, Futo, Ito-shi

都心からのアクセスが抜群なダイビングスポットとして、長年多くのダイバーに愛されている。ダイビングだけでなく、夏期限定でオープンする「日本一海に近い磯プール」も人気で、海水浴気分でプールを楽しむことができる。

## Izu Oceanic Park

It has long been loved by many divers as a diving spot with its excellent access from the center of Tokyo. "Iso Pool (the seashore pool)," which is the closest swimming pool to the sea in Japan and is open only in summer, is also popular. You can swim in the pool as if you were enjoying sea bathing.

### 旅酒が見つかる場所 Shop

● ルネッサ赤沢 売店
伊東市赤沢190-5
0557-54-1670

※最新の販売店情報はホームページでご確認ください。
Please check our website for the latest retail shop information.

39

# TOTTORI SAND DUNES -Tottori-

# 鳥取砂丘 (鳥取県)

日本海海岸に広がる広大な砂礫地で、日本を代表する海岸砂丘である鳥取砂丘。圧倒的な存在感、世界観、広大な景色は訪れる人に感動を与えてくれます。ラクダに乗ったり、砂地を走るファットバイクに乗ったり、アウトドアアクティビティも充実しており、景色を堪能するだけでなく、さまざまな魅力を持つ観光地です。

The Tottori Sand Dunes is a huge gravely sand lying by the coastline of the Sea of Japan. The presence of the sand is overwhelming and the scenery impresses many visitors. You can enjoy a camel ride, a fat bike ride on sandy ground and other outdoor activities, as well as its beautiful scenery. It is a tourist spot with various attractions.

## 旅 TABI-SAKE 酒

### TOTTORI SAND DUNES

鳥取県固有の幻の酒米「強力（ごうりき）」を50%精米した純米大吟醸酒です。穏やかな吟醸香とシャキッとした味わいで食中酒としてお楽しみいただけます。アゴ（飛魚）竹輪、とうふ竹輪などの郷土料理との相性は抜群です。

It is the Junmai Daiginjo-shu using 50% of the "Goriki," which is a special sake rice in Tottori prefecture. It has a refreshing taste and is best while eating. It goes well with cultural foods such as flying fish, chikuwa, and Tofu chikuwa.

**醸造元：**中川酒造株式会社
（1828年創業／鳥取県）
**種類：**日本酒 純米大吟醸酒
**度数：**15度
**精米歩合：**50%

**Brewery :** Nakagawa Shuzo Co., Ltd.
（Since 1828／Tottori）
**Kind :** Junmai Daiginjo-shu
**Alcohol Content :** 15%
**Polishing Ratio :** 50%

**甘辛度**
Sweet-dryness

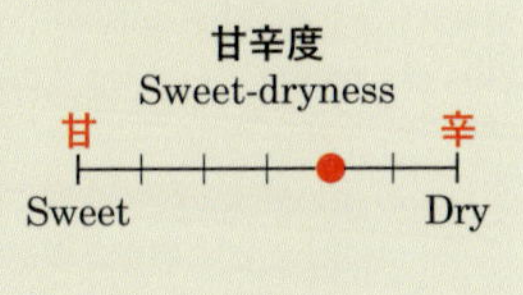

## 主な行き方 Major Directions

羽田空港＝(75分)⇒鳥取空港／バス＝(20分)⇒鳥取駅／バス＝(25分)⇒みどり町駅／徒歩＝(15分)⇒鳥取砂丘

Haneda Airport＝(75 min)⇒Tottori Airport／Bus＝(20 min)⇒Tottori Station／Bus＝(25 min)⇒Midorimachi Station／Walking＝(15 min)⇒Tottori Sand Dunes

### ▶①砂の美術館

世界初の砂像を専門に展示する美術館。世界でも最高レベルの砂像を展示しています。

### ▶②砂丘温泉ふれあい会館

鳥取砂丘で遊んで砂まみれになったら訪れておきたい砂丘温泉。鳥取砂丘の近くにあり、大きなガラス越しに日本海と鳥取砂丘を眺めながらお湯につかることができます。JR鳥取駅からはバスで25分程のところにあります。

### ▶①The Sand Museum

It is the world's first Museum displaying sand statues. The Museum has one of the best sand statues in the world.

### ▶②Sakyu Onsen Fureai Hall

The Sakyu Onsen Fureai Hall is a place where you want to visit after playing in the sand. It is near the Tottori Sand Dunes and you can take a bath while seeing the Japanese Ocean and the desert from a large window. It is located at a place 25 min in bus from the JR Tottori Station.

**Brewery's recommended spot**
**蔵元おすすめスポット**

### 鳥取砂丘

鳥取市福部町湯山2164-661
2164-661, Yuyama, Fukube-cho, Tottori-shi

夕日と砂丘、星空と砂丘、雪と砂丘など、四季や時間によって刻々と姿を変えてゆく様が魅力的。

### Tottori Sand Dunes

The sunset and desert, the stars and the desert, the snow and the desert; it changes its feature as the season changes.

フレンチルネッサンス様式の西洋館。白亜の建物と日本庭園の美しさを見るために多くの観光客が訪れる。

## 仁風閣

鳥取市東町2-121

2-121, Higashimachi, Tottori-shi

## Jinpukaku

The European Style House of the French Renaissance. To see the chalk white building and the Japanese garden, many people visit here.

## 麒麟獅子舞

因幡・但馬の地域に伝わる民俗芸能。地域のさまざまな祭礼行事等で年間通して舞われている。

## Kirinjishi-mai

A folk entertainment of the Tajima and Inaba region. It is danced almost all the time during the year for all sorts of events of the region.

落差40mの滝で迫力満点。
周囲にも大小48もの滝が存在し、いずれも自然のままの滝を見ることができる。

## 雨滝

鳥取市国府町雨滝

Amedaki, Kokufu-cho, Tottori-shi

## Amedaki Waterhall

It has about a fall of 40m. Around the waterfall, there are about 48 other big and small waterfalls, and you can see all of them naturally.

### 旅酒が見つかる場所 Shop

- **アベ鳥取堂 鳥取駅南売店**
  鳥取市東品治町111-1
  JR鳥取駅
  080-2924-2793
- **鳥取大丸**
  鳥取市今町2-151 地下1階
  0857-25-2111 ㈹
- **道の駅 清流茶屋かわはら 物産館 鮎遊座**
  鳥取市河原町高福837
  0858-85-6205
- **鳥取砂丘会館**
  鳥取市福部町湯山2164
  0857-22-6835
- **砂丘センター**
  鳥取市福部町湯山2083
  0857-22-2111
- **らくだや**
  鳥取市福部町湯山2164-806
  0857-23-1735

※最新の販売店情報はホームページでご確認ください。
Please check our website for the latest retail shop information.

40

## OSAKA KITA・MINAMI -Osaka-

# 大阪キタ・ミナミ（大阪府）

西日本最大の都市大阪。梅田周辺をキタと呼び、難波周辺をミナミと呼びます。特徴の違うふたつのエリア、ともに味わってみてはいかがでしょうか。

Osaka is the largest city in West Japan. We call the area around Umeda "Kita" and the area around Namba "Minami." Please enjoy two different parts of Japan.

### 旅 TABI-SAKE 酒

**OSAKA KITA ・ MINAMI**

当蔵が得意としている9号酵母による味わい重視の純米吟醸です。香りを抑え、米の旨みを引き立たせた味わいは、香り酒の苦手な方や、酒好きの方に好評をいただいています。日常に楽しむ高級酒としても、パーティや宴席などの公式の場での格式を持ったお酒としても、自信を持っておすすめします。冷やしても燗でもOK。

This Special Junmai-shu is made by using the No.9 yeast which the brewery is great at making and the sake is focused on its taste. The scent is restricted, so the taste of the rice is enlivened; people who the smell of alcohol and people who like alcohol both favor this one. We highly encourage this sake as a classy alcohol to enjoy daily, and for parties and ceremonies. It goes well with cold or hot.

**醸造元：**壽酒造株式会社
（1822年創業／大阪府）
**種類：**日本酒 特別純米酒
**度数：**16%
**精米歩合：**60%

**Brewery** : Kotobuki Brewing Co., Ltd.
(Since 1822／Osaka)
**Kind** : Special Junmai-shu
**Alcohol Content** : 16%
**Polishing Ratio** : 60%

**甘辛度**
Sweet-dryness

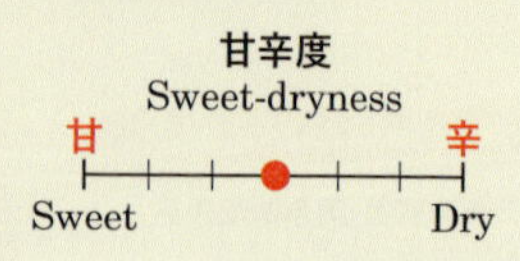

## 主な行き方 Major Directions

大阪国際空港＝(2分)⇒蛍池駅＝(15分)⇒阪急梅田駅＝(10分)⇒梅田駅＝(10分)⇒難波駅

Osaka International Airport＝(2 min)⇒Hotaruike Station＝(15 min)⇒Hankyu Umeda Station＝(10 min)⇒Namba Station

## 旅の宝 Treasure Hunt

### ▶①梅田スカイビル

大阪キタの代表的な超高層ビル。最上部の空中庭園とその独自の設計から、世界的に高く評価されている建造物で、海外からも多くの観光客が訪れています。

### ▶②通天閣

難波のシンボル通天閣。地元民に愛され続けるこのタワー。他のタワー程高くはありませんが、高さではなく、「面白さ」がその人気のポイントです

### ▶③あべのハルカス

大阪の新しい観光名所で近年多くの観光客を集めているあべのハルカス。地上300mの日本一高いビルで、展望台からは大阪の街が一望できます。時間帯によって景色が変わり、夜になると、通天閣がふたつに見えるポイントもあります。

### ▶①Umeda Sky Building

It is a tower building representing Kita Osaka. At the top of the building, there is a hanging garden and the design of the building is evaluated highly.it is a popular building for many tourists from outside of Japan.

### ▶②Tsutenkaku Tower

Tsutenkaku Tower is the symbol of Namba. It is favored by the local people. It is not as high as the other towers but it is popular for its feature.

### ▶③Abeno Harukas

Abeno Harukas is Osaka's new tourist attraction and has attracted many tourists in recent years. It is the tallest building in Japan, 300 meters above ground, and you can see the whole city of Osaka from the observation deck. The view from there changes depending on the time of day, and there is a point where you can experience a strange phenomenon that looks like there are two Tsutenkaku Towers at night.

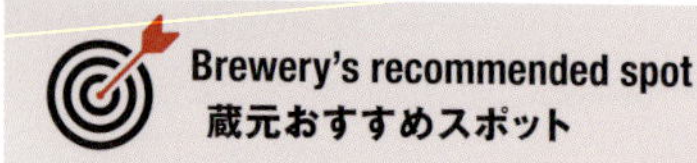


## 水掛不動尊(法善寺)

大阪市中央区難波1-2-16
1-2-16, Namba, Chuou-ku, Osaka-shi

法善寺はミナミの繁華街にある寺院。
水掛不動尊には参拝者が後を絶たない。
水掛不動尊の不動明王は苔に覆われている姿で有名。
風情ある石畳に飲食店が立ち並ぶ法善寺横丁も多くの人でにぎわっている。

## Mizukake Fudoson (Hozenji Temple)

Temple at the busy street of Minami. There are always people visiting the Mizukakefudoson. The Fudo is famous for it being covered with moss. There are also many people at the Hozenji Yokocho, where many restaurants are standing on the stone pavement.

**旅酒が見つかる場所** **Shop**

- **ビヤレストラン ニユートーキヨー 第一生命ビル店**
  大阪市北区梅田1-8-17
  第一生命ビル 地下2階
  06-6346-7451

- **河岸番外地 堂島店**
  大阪市北区堂島2-2-2
  近鉄堂島ビル地下1階
  06-6345-4380

※最新の販売店情報はホームページでご確認ください。
Please check our website for the latest retail shop information.

# SATSUMA -Kagoshima-

41

# 薩摩 (鹿児島県)

鹿児島県の西半分、古くから薩摩国として栄えて来た薩摩エリア。自然資源豊富なこの地域は自然を利用した観光スポットが数多くあります。明治維新に活躍した偉人も多く、歴史的な名所もあり見所いっぱいのエリアです。

The Satsuma region occupies the western half of Kagoshima prefecture and has prospered as Satsuma Province since ancient times. This area is rich in natural resources and has many tourist spots that take advantage of nature. It has produced many great people who played active roles in the Meiji Restoration, and is full of attractions including famous historical sights.

## 旅 TABI-SAKE 酒 SATSUMA

鹿児島県産のコガネセンガン、米麹に国産米と原材料には確かな品質のものを選び、丁寧に造られた逸品。白麹ならではのスッキリとした飲み口で、お料理との相性が良い酒質です。

The shochu is made from using Japanese made malted rice koganesengan from Kagoshima prefecture and takes special care to its ingredients. The aftertaste is very refreshing and it goes well with any type of food.

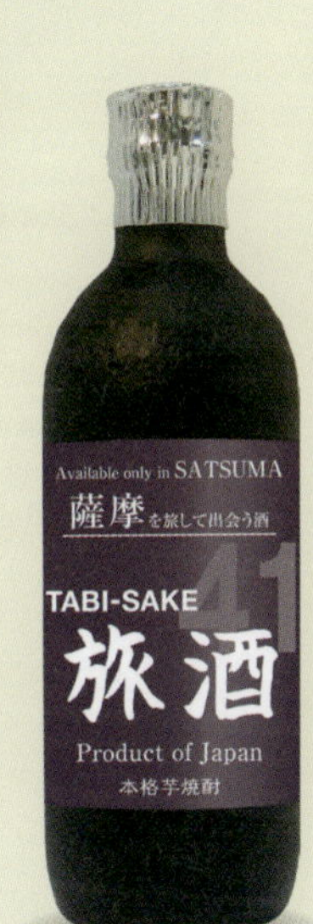

**醸造元：**大口酒造株式会社
（1970年創業／鹿児島県）
**種類：**本格芋焼酎
**度数：**25%

**Distillery** : Okuchi Shuzo Co., Ltd.
（Since 1970／Kagoshima）
**Kind** : Honkaku-shochu
**Alcohol Content** : 25%

## 主な行き方 Major Directions

羽田空港＝(110分)⇒鹿児島空港／バス＝(40分)⇒鹿児島中央駅
Haneda Airport＝(110 min)⇒Kagoshima Airport／Bus＝(40 min)⇒Kagoshima Central Station

## 旅の宝 Treasure Hunt

### ▶①竜石

人間に目撃された女竜が石に変わったという伝説のある岩。恋愛成就のパワースポットとしても人気のこの岩は、初心者でも楽しめる登山コースでみつけることができます。

### ▶②藺牟田池

飯盛山の噴火によりできたとされる火口湖。泥炭形成植物群落でできた浮島は、2005年にラムサール条約指定湿地に登録されています。

### ▶③桜島

鹿児島といえば桜島。大正噴火によって流れ出た溶岩の上にある観光施設「有村溶岩展望所」の遊歩道からは、桜島や錦江湾を一望することができる。

### ▶①Tatsuishi

This rock has a folk story that a female dragon turned in to this rock because a human being saw her. This rock is known for success in love and is located on a hiking way which beginners could enjoy it.

### ▶②Imuta Pond

This crater lake was made due to the evaporation of Mt.Iimori. In 2005, it was register the wetlands of Imuta Pond are registered under the Ramsar Convention.

### ▶③Sakurajima Island

Kagoshima prefecture is known for Sakurajima Island. The Arimura Lava Observation spot is built on the lava from the evaporation of Taisho, and from its promenade you can see the view of Sakurajima Island and Kinko Bay.

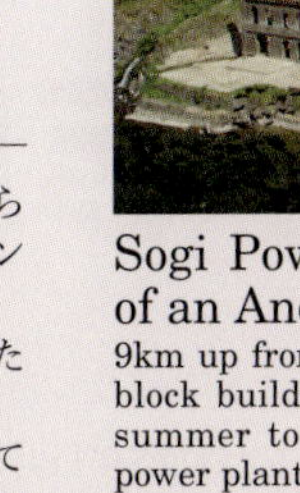

Distillery's recommended spot
蔵元おすすめスポット

## 曽木発電所遺構

伊佐市大口宮人　Miyahito, Okuchi, Isa-shi

鶴田ダムの上流9kmの場所に初夏から秋にかけて、期間限定で姿を現す赤レンガ造りの建物。
明治時代、牛尾大口金山の電源供給のために建造された水力発電所の跡。
冬から春の間は、建物上部だけを残して水に沈んでいる。

## Sogi Power Station Remains of an Ancient Structure

9km up from the Tsuruta Dam, the red block building shows itself from early summer to autumn. It is a hydraulic power plant made to generate electricity for the Ushio-okuchi Mine. Through winter to spring, it is under water.

## 霧島

霧島市　Kirishima-shi

霧島の中央にある霧島神宮など、霧島も鹿児島を訪れるならば一度はまわってみたい観光地。標高600～850mの場所にある霧島温泉郷は、眺望自慢の宿も多く、日頃の疲れを癒すには最適。

## Kirishima

You should visit Kirishima once when you have a chance to visit Kagoshima prefecture. It has many tour spots such as the Kirishimajingu Shrine. The Kirishima Onsen Village is located at a place 600-850 m high and many hotels have amazing views; it is a great place to relax and relieve your stress.

## 指宿

指宿市　Ibusuki-shi

天然砂むし温泉（砂風呂）で有名な指宿。砂風呂はデトックス効果が高いといわれ、女性からの人気が高い。近隣の知覧には、武家屋敷庭園群や知覧特攻平和会館など、歴史的なスポットも多く存在する。

## Ibusuki

It is famous for sand bath. Sand bath has a high detoxification and it is popular among women. In Chiran, there are famous historical places such as the old samurai residences garden village, and the Chiran Peace Museum.

**旅酒が見つかる場所**　**Shop**

**●曾木の滝 なりざわ**
伊佐市大口宮人635-11
曾木の滝公園内
0995-28-2345

**●ドルフィンポート山形屋店 薩摩酒蔵**
鹿児島市本港新町5-4
099-221-5823

※最新の販売店情報はホームページでご確認ください。
Please check our website for the latest retail shop information.

42

## ISESHIMA -Mie-

# 伊勢志摩 (三重県)

国内神社の最高峰、伊勢神宮を中心とした観光地、伊勢志摩。リアス式海岸として有名な英虞湾の美しさを目に焼き付けるのもよし、海を生かしたテーマパークで楽しむのもよし、伊勢神宮周辺の昔ながらの情緒が残る町並みを楽しみながら歩くのもよし、と、伊勢志摩は人々を魅了するスポットの宝庫です。

Iseshima is known for the Shinto's most sacred shrine in Japan, the Isejingu Shrine. It is a treasure trove of attractions that captivates many visitors. Some might want to see the beauty of Ago Bay, which is famous for its ria coastline; some might want to have a joyful time at a theme park, which features the sea; or others might like taking a walk while enjoying the traditional townscape around the Isejingu Shrine.

## 旅酒 TABI-SAKE ISESHIMA

まろやかな旨みと調和を楽しめる酒。

A sake that has a harmony with a mild umami.

**醸造元**：河武醸造株式会社
（1857年創業／三重県）
**種類**：日本酒 純米酒
**度数**：15度
**精米歩合**：65%

**Brewery** : Kawatake Jhouzo Co., Ltd.
(Since 1857／Mie)
**Kind** : Junmai-shu
**Alcohol Content** : 15%
**Polishing Ratio** : 65%

甘辛度
Sweet-dryness
甘 Sweet — 辛 Dry

## 主な行き方 Major Directions

羽田空港＝(20分)⇒品川駅＝(95分)⇒名古屋駅＝(90分)⇒伊勢市駅＝(35分)⇒志摩磯部駅

Haneda Airport＝(20 min)⇒Shinagawa Station＝(95 min)⇒Nagoya Station＝(90 min)⇒Ise-shi Station＝(35 min)⇒Shima Isobe Station

### ▶①伊勢神宮

天照大御神を奉る国内神社の最高峰伊勢神宮。2000年の歴史を有し、内宮・外宮をはじめ125の宮社が観光客を迎えてくれます。日本神話を知るなら、一度は訪れておきたい観光スポットです。

### ▶①Isejingu Shrine

The Isejingu Shrine is the greatest shrine in Japan which offers to the Amaterasu Omikami. It has about 2000 years of history and has 125 Gu-titled shrines. If you want to know the Japanese mythology, it a place where you should visit once.

### ▶②夫婦岩

大注連縄で太く堅く結ばれた2つの岩（男岩、女岩）。夫婦やカップルなど、恋愛の象徴とされている縁結びのシンボルとしても有名です。

### ▶②Meoto Iwa (Married Rocks)

There are two rocks been tied up with a large sacred rope. This place is famous for its success in love. Many couples visit this spot for their matchmakings.

## Brewery's recommended spot
## 蔵元おすすめスポット

### 伊勢忍者キングダム

伊勢市二見町三津1201-1

1201-1, Mitsu, Futami-cho, Ise-shi

今から400年前の町並みが再現され、当時の文化や人々の暮らしを肌で感じることができるテーマパーク。
2018年に新登場したアトラクション「忍者森のアドベンチャー」は子どもたちに人気。

### Ninja Kingdom Ise

The theme park is replicated from itself 400 years ago, and you can enjoy the atmosphere of the past. The children love the attraction, "Ninjamori Adventure," built in 2018.

### 大王崎灯台

志摩市大王町波切54

54, Nakiri, Daio-cho, Shima-shi

志摩半島の東南端にある大王崎の先端に立つ白亜の灯台で参観可能な灯台。

### Daiozaki Lighthouse

The lighthouse is located at the east south of the Ise island and people can take a look inside.

## 旅酒が見つかる場所 Shop

- **酒のあおき**
  伊勢市宮町2-2-1
  0596-28-2979
- **岡七酒店**
  伊勢市宇治中之切町4
  0596-22-4586
- **髙橋酒店**
  伊勢市宇治今在家町43
  0596-22-2567
- **奥志摩の酒商人べんのや**
  志摩市志摩町和具3065
  0599-85-0420

※最新の販売店情報はホームページでご確認ください。
Please check our website for the latest retail shop information.

# KAGA・KANAZAWA -Ishikawa-

# 加賀・金沢 (石川県)

43

「加賀百万石」の城下町として栄えた金沢。金沢市内には、歴史的に高い価値を持つ建造物が多く残り、絢爛豪華な加賀の歴史を垣間見ることができます。北陸三県にまたがる加賀温泉郷など自然を活かした観光地もあり、多くの観光客を惹きつけているエリアです。

Kanazawa is a castle town that prospered as the "fief of Kaga." Many traditional buildings with high historical value remain in Kanazawa-shi, and you can get a glimpse of the gorgeous history of Kaga (the old name of Kanazawa). Its surrounding area also attracts a lot of visitors since there are tourist spots that take advantage of nature such as Kaga Onsenkyo, which straddles three prefectures in Hokuriku.

## 旅 TABI-SAKE 酒

KAGA・KANAZAWA

スッキリとした酸味のなかに、深い味わいのある純米酒。

The Junmai-shu that has a deep taste inside the refreshing sourness.

**醸造元**：株式会社加越
(1865年創業／石川県)
**種類**：日本酒 純米酒
**度数**：14.5%
**精米歩合**：75%

**Brewery**：Kaetsu Co., Ltd. (Since 1865／Ishikawa)
**Kind**：Junmai-shu
**Alcohol Content**：14.5%
**Polishing Ratio**：75%

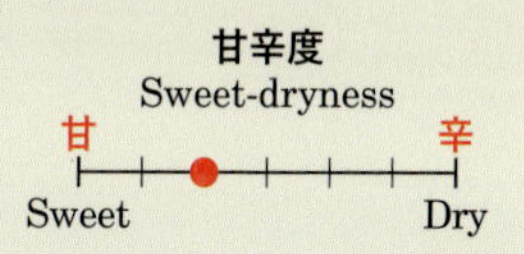

## 主な行き方 Major Directions

羽田空港＝(60分)⇒小松空港／バス＝(40分)⇒金沢駅

Haneda Airport＝(60 min)Komatsu Airport／Bus＝(40 min)⇒Kanazawa Station

## 旅の宝 Treasure Hunt

### ▶①兼六園

日本三名園のひとつ。国の特別名勝にも指定されている日本庭園です。桜や紅葉、雪吊りと四季折々の美しさが楽しめます。

### ▶①Kenrokuen Garden

It is one of the three famous Japanese gardens. It is also registered as the country's special scenic beauty. You can enjoy the beauty of each season such as sakura, red leaves, and snow.

### ▶②長町武家屋敷跡

加賀藩時代の上流・中流階級藩士の侍屋敷が軒を連ねている長町武家屋敷跡。土塀と石畳の路地が続いており、藩政時代の情緒ある雰囲気を味わうことができます。

### ▶②Nagamachi Old Samurai Residences

Nagamachi Old Samurai Residences are houses for the upper and middle feudal retainers during the Kagahan period. You can enjoy the atmosphere from the feudal government period from the mud walls and stone pavements.

**Brewery's recommended spot**
**蔵元おすすめスポット**

## ひがし茶屋街

金沢市東山
Higashiyama, Kanawawa-shi

江戸時代に建てられたお茶屋が中心にあり、有料で内部を見学することができる。和菓子店や土産物店などが多くの店が立ち並び、多くの観光客でにぎわっている。

## Higashi Chaya District

The geisha houses in the Edo period are at the middle of the town and you can tour inside. Many tourists visit for souvenirs and wagashi at the place.

## 金沢城

金沢市丸の内1-1
1-1, Marunouchi, Kanazawa-shi

加賀百万石として、江戸時代に町の中心としてにぎわっていた金沢城。
江戸時代から残る石川門は特に有名。白い漆喰が塗られている「なまこ壁」が城門の美しさの秘訣である。

## Kanazawa Castle

Kanazawa Castle was the center of the town during the Edo period. The Ishikawa gate is very famous, still standing from the Edo period; the white painting of the wall, known as "Namako Wall," is very beautiful.

**旅酒が見つかる場所** **Shop**

- **酒の井なみ屋**
  金沢市出雲町イ45ー2
  076-233-1738
- **加賀 山代温泉 ゆのくに天祥**
  加賀市山代温泉19-49-1
  0761-77-1234
- **ひがしやま酒楽**
  金沢市東山1-25-5
  076-251-1139

※最新の販売店情報はホームページでご確認ください。
Please check our website for the latest retail shop information.

44

## SETO OHASHI -Kagawa-

# 瀬戸大橋（香川県）

四国と本州を結ぶ瀬戸大橋。海峡部9.4kmに架かる6橋を総称して瀬戸大橋と呼ばれています。世界最大級の橋梁が連なる姿は壮観です。車で渡ったり、電車で渡ったり、クルージングで橋を下から見あげたり、時間帯によって日の出や日の入りを見たり、ライトアップされた姿を見たり、と楽しみ方はさまざまあります。

Seto Ohashi Bridge is the bridge that connects the main land and Shikoku region. The bridge is structured by 6 parts and has a length of 9.4 km. The view of the largest bridge in the world is overwhelming.
You can cross the bridge by car, by train, or look it up from below while enjoying cruising. There are various ways to enjoy the bridge: the sunrise and sunset seen from the bridge is beautiful, and its illuminated figure at night is spectacular.

### 旅酒 TABI-SAKE
SETO OHASHI

まろやかさ・旨みのある旨口純米酒。香川県産の酒米「オオセト」100%使用。このお酒はコクのある味つけや旨みのある食材と好相性です。

Junmai-shu with mild and umami taste. Using 100% of "Oseto" rice from Kagawa prefecture. This sake goes well with foods that have koku and umami.

**醸造元**：西野金陵株式会社
（1789年創業／香川県）
**種類**：日本酒 純米酒
**度数**：15度以上16度未満
**精米歩合**：70%

**Brewery**：Nisinokinryo Corp.（Since 1789／Kagawa）
**Kind**：Junmai-shu
**Alcohol Content**：15-16%
**Polishing Ratio**：70%

甘辛度
Sweet-dryness
甘 Sweet — 辛 Dry

## 主な行き方 Major Directions

羽田空港＝(80分)⇒高松空港／バス＝(45分)⇒高松駅＝(15分)⇒坂出駅／徒歩＝(80分)⇒瀬戸大橋

Haneda Airport＝(80 min)⇒Takamatsu Airport／Bus＝(45 min)⇒Takamatsu Station＝(15 min)⇒Sakaide Station／Walking＝(80 min)⇒Seto Ohashi Bridge

### ▶①瀬戸大橋

2018年に開通30周年を迎える瀬戸大橋。瀬戸内海の壮大な自然のなか、ライトアップされた瀬戸大橋は非常に美しく多くの観光客を魅了しています。

### ▶①Seto Ohashi Bridge

In 2018, the bridge welcomes its 30th anniversary. In the nature of Setnai sea, the bridge being lighten up is beautiful and amazes many tourists.

### ▶②瀬戸大橋タワー（瀬戸大橋記念公園）

瀬戸大橋記念公園内にあるタワー。地上108mから、瀬戸大橋や瀬戸内海の島々を眺めることができます。

### ▶②Seto Ohashi Bridge Tower (Seto Ohashi Bridge Memorial Park)

The tower in the park is called the Seto Ohashi tower. You can see the view of Seto Ohashi Bridge and the islands in the Setonai sea from the top of the tower which is 108 m from the ground.

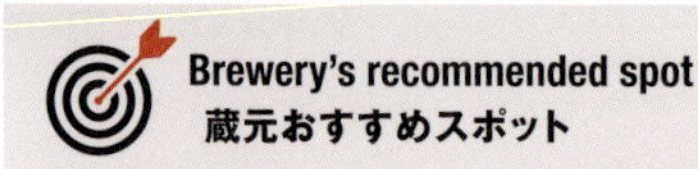

## 金刀比羅宮

仲多度郡琴平町892-1
892-1, Kotohira-cho Nakatado-gun

象頭山の中腹に鎮座する「さぬきこんぴらさん」。御本宮まで続く長い石段は785段ある。
古くから信仰の地となっており、江戸時代には「こんぴら参り」が全国に広がっていた。

## Kotohiragu Shrine

"Sanuki Kompirasan" lies at the middle of Mt. Zozusan. Till the shrine, there are about 785 stairs. It is a sacred place and the "kompira-mairi" is a well known for activity all over the coutry.

## 栗林公園

高松市栗林町1-20-16
1-20-16, Ritsurin-cho, Takamatsu-shi

高松藩主松平家の別邸として、300年近く前に完成した公園。1000本もの見事な手入れ松をはじめ、四季折々の花々を見ることができる。香川県の梅の開花標本木もここにある。

## Ritsurin Garden

This garden was completed nearly 300 years ago as the second residence of the Matsudaira family, the lord of the Takamatsu Domain. You can see 1000 beautifully pruned pine trees and seasonal flowers in the park. Here is a plum tree that is the standard for the flowering of plum blossoms in Kagawa prefecture.

**旅酒が見つかる場所** Shop

**四国ショップ88**
高松市サンポート2-1
高松シンボルタワー1階
087-822-0459

※最新の販売店情報はホームページでご確認ください。
Please check our website for the latest retail shop information.

45

# TATEYAMA KUROBE UNAZUKI -Toyama-

# 立山黒部宇奈月（富山県）

北陸の豪雪地帯、立山エリア。雄大な立山連峰は日本有数の山岳観光スポット。高さ日本一の黒部ダムやトロッコ電車で訪れる宇奈月温泉など、豊かな自然からなる観光スポットが盛りだくさんです。

The Tateyama area is a heavy snowing region located in the northern part of Japan. The Tateyama mountains have one of the best mountain tourist spots. There are many touring spots such as the Kurobe Dam, the highest damn in Japan, and the Unazuki Onsen, where you can visit by using the trolley.

## 旅 TABI-SAKE 酒

### TATEYAMA KUROBE UNAZUKI

食事が楽しくなるクセのない飲みやすい辛口です。

The dry taste of the Junmai-shu makes you enjoy your food.

**醸造元**：銀盤酒造株式会社（1910年創業／富山県）
**種類**：日本酒 純米酒
**度数**：16度
**精米歩合**：60%

**Brewery** : Ginban Shuzo Co., Ltd. (Since 1910／Toyama)
**Kind** : Junmai-shu
**Alcohol Content** : 16%
**Polishing Ratio** : 60%

**甘辛度**
Sweet-dryness

甘 Sweet ——●— 辛 Dry

## 主な行き方 Major Directions

東京駅＝(新幹線160分)⇒黒部宇奈月温泉駅／徒歩＝(10分)⇒新黒部駅＝(30分)⇒宇奈月温泉駅

Tokyo Station＝(Shinkansen160 min)⇒Kurobe-Unazukionsen Station／Walking＝(10 min)⇒Shin-Kurobe Station＝(30 min)⇒Unazukionsen Station

東京駅＝(新幹線140分)⇒富山駅／徒歩＝(10分)⇒電鉄富山駅＝(70分)⇒立山駅／ケーブルカー＝(7分)⇒美女平／バス＝(50分)⇒室堂／トロリーバス＝(10分)⇒大観峰／ロープウェイ＝(7分)⇒黒部平／ケーブルカー＝(5分)⇒黒部湖／徒歩＝(10分)⇒黒部ダム

Tokyo Station＝(Shinkansen140 min)⇒Toyama Station／Walking(10 min)⇒Dentetsu-Toyama Station＝(70 min)⇒Tateyama Station／Cable car＝(7 min)⇒Bijodaira／Bus＝(50 min)⇒Murodo／Trolley bus＝(10 min)⇒Daikanbo／Ropeway＝(7 min)⇒Kurobe Daira／Cable car＝(5 min)⇒Lake Kurobe／Walking＝(10 min)⇒Kurobe Dam

## 旅の宝 Treasure Hunt

### ▶①宇奈月温泉

黒部川の渓谷沿い、トロッコ電車で巡る宇奈月温泉。美肌の湯として有名で、全国から観光客が集まる名湯です。

### ▶②黒部ダム

黒部川に建設された水力発電専用のダム。6月下旬から10月中旬に行われているダムからの放水は迫力満点。ダム内部の見学もできます。

### ▶①Unazuki Onsen

The Unazuki Onsen is located along the valley of Kurobe River where people ride the trolley to get there. The hot spring is famous for having a great effect on the skin and many tourists, all over of Japan, visit here.

### ▶②Kurobe Dam

The Kurobe Dam is built on the Kurobe River to generate electricity. From late June to October, the Dam releases the water. Also, people are able to tour inside the dam.

**Brewery's recommended spot**
**蔵元おすすめスポット**

## 黒部の放水

中新川郡立山町芦峅寺
Ashikuraji, Tateyama-machi, Nakaniikawa-gun

黒部ダム名物である観光放水は毎年6月下旬～10月中旬に実施される。
観る場所によって印象が異なるので、それぞれのスポットからその違いを実感できる。
大迫力の放水を背景に写真を撮ることができる記念写真撮影システムもある。

## Water Discharge at Kurobe Dam

The release of the water of Kurobe Dam is held from late June to middle of October. The view of it is very different at each spot. There are memorial photo booths where you can take pictures with the water being released.

## 旅酒が見つかる場所 Shop

- **魚の駅「生地」**
  黒部市生地中区265
  0765-57-0192
- **道の駅うなづき「うなづき食彩館」**
  黒部市宇奈月町下立687
  0765-65-2277
- **米沢商店**
  黒部市宇奈月温泉384
  0765-62-1159
- **黒部市地域観光ギャラリー「のわまーと」**
  黒部市若栗3212-1
  0765-57-2853

※最新の販売店情報はホームページでご確認ください。
Please check our website for the latest retail shop information.

**◆他にも素敵な場所があります。あなただけの旅の宝を探してみましょう!**
**There are other wonderful places to visit. Let's take a trip to find a treasure of your own!**

46

# 京都市 （京都府）

かつての日本の首都、京都。皇室の宮殿や御所、神社、寺院はもちろん、伝統的な木造建築から史跡など、日本の歴史と文化が詰まった日本を代表する観光名所です。多くの観光スポットがあり、また紅葉の季節や桜の季節など、訪れる時期限定の絶景も多いため、何度も訪れる人が多い観光地でもあります。

Kyoto used to be the capital of Japan. Kyoto is one of the best touring spots in Japan full of Japanese cultures and histories such as the emperor's palace, shrines, and many other buildings made with wood. The city has a lot of tourist attractions, with each season featuring the seasonal beauty such as cherry blossoms in spring and colored leaves in autumn. This is the place people visit repeatedly.

## 旅 TABI-SAKE 酒

KYOTO CITY

大吟醸特有のフルーティーな吟醸香が香る、スッキリとした味わいのお酒です。京都府産「京の輝き」米を100%使用し、原料も京都にこだわったお酒です。

The Junmai Daiginjo-shu is refreshing since it uses top-quality sake brewed from rice grains. Using 100% of "Kyou no Kagayaki" from Kyoto, and all the ingredients are related with Kyoto.

**醸造元：** 齊藤酒造株式会社
（1895年創業／京都府）
**種類：** 日本酒 純米大吟醸酒
**度数：** 13度
**精米歩合：** 50%

**Brewery** : Saito Shuzo Co., Ltd.
（Since 1895／Kyoto）
**Kind** : Junmai Daiginjo-shu
**Alcohol Content** : 13%
**Polishing Ratio** : 50%

甘辛度
Sweet-dryness

## 主な行き方 Major Directions

羽田空港＝(20分)⇒品川駅＝(130分)⇒京都駅

Haneda Airport＝(20 min)⇒Shinagawa Station＝(130 min)⇒Kyoto Station

## 旅の宝 Treasure Hunt

### ▶①清水寺

平安京遷都以前からの歴史を持つ、有名な「舞台」のある古寺です。古都京都の文化財としてユネスコ世界遺産に登録されています。

### ▶①Kiyomizudera Temple

The Kiyomizudera Temple has lasted before the Heian period, and is famous for having the stage of Kiyomizu. It is also registered as the World Heritage Site.

### ▶②金閣寺

京都観光の目玉の一つにも上げられる金閣寺。正式には鹿苑寺といいます。建物の内外に金箔を貼った舎利殿が池に反射された風景の美しさは必見です。

### ▶②Kinkakuji Temple

The Kinkakuji Temple is a must-see touring spot in Kyoto. The formal name is Rokuonji Temple. The building's surface is laminated with gold and the view of the building reflecting on the lake is glorious.

Brewery's recommended spot
蔵元おすすめスポット

## 伏見稲荷

京都市伏見区深草藪之内町68
68, Yabunouchi-cho, Fukakusa, Fushimi-ku Kyoto-shi

全国に30,000社あるといわれる「お稲荷さん」の総本宮が伏見稲荷大社。
朱塗りの鳥居が連なる千本鳥居でお馴染み。

## Fushimiinari Shrine

The main shrine of "Oinaisan", having about 30,000 of them in Japan. Many people have seen the red gates being aligned.

東福寺の塔頭で、庭の苔がとても美しいことから「虹の苔寺」とも呼ばれる。白砂の枯れ池と苔を巧みに調和させた枯山水庭園「波心庭」があり、春は桜やサツキ、ツツジ等の色とりどりの花、夏は新緑、秋は紅葉、四季折々の美しい風景を見ることができる。

## 光明院

京都市東山区本町15-809
15-809, Honmachi, Higashiyama-ku Kyoto-shi

## Komyoin Temple

The building of Tofukuji Temple, and the garden is covered with moss and is also called as the rainbow moss shrine. The white sand, the pond, and the moss of the garden are well harmonized in the Hashintei Garden. You can enjoy many types of flower blooming in each season.

### 旅酒が見つかる場所 Shop

- **くらま アスティ京都店**
  京都市下京区東塩小路高倉町8-3 京都八条口アスティロードレストラン街
  075-681-1539
- **京土愛**
  京都市下京区烏丸塩小路北面角 京都タワーサンド 1階
  075-353-2499
- **京都ハンディクラフトセンター**
  京都市左京区聖護院円頓美町17
  075-761-8001
- **卯田光商店**
  京都市東山区東大路上田町85
  075-561-4173

※最新の販売店情報はホームページでご確認ください。
Please check our website for the latest retail shop information.

# SENDAI -Miyagi-

# 仙台 (宮城県)

47

東北地方最大の都市仙台。自然が豊富にあることから、杜（もり）の都とも呼ばれます。都市部と自然が見事に調和するこの街は、グルメから史跡巡りまで、さまざまな楽しみ方ができます。

Sendai is the largest city in the Tohoku area. Since it is rich in nature, it is also called as the city of forest. People can enjoy sightseeing the Sendai, rich in nature and having a downtown, in many different ways such as eating good food and visiting historical landmarks.

## 旅 TABI-SAKE 酒

**SENDAI**

フルーティーな香りと濃厚な酸味、そしてのどごしに感じるアルコールの刺激。それはまるで完熟の果物をまるごと搾ったような味わいです。

The fruity fragrance and the thick sourness is very stimulating as if you have squeezed out a fruit.

**醸造元**：森民酒造本家
（1849年創業／宮城県）
**種類**：日本酒 特別純米酒
**度数**：16度
**精米歩合**：60%

**Brewery** : joint-stock company Morin people Sake brewery（Since 1849／Miyagi）
**Kind** : Special Junmai-shu
**Alcohol Content** : 16%
**Polishing Ratio** : 60%

## 主な行き方 Major Directions

羽田空港＝(20分)⇒浜松町駅＝(40分)⇒東京駅＝(95分)⇒仙台駅
Haneda Airport＝(20 min)⇒Hamamatsucho Station＝(40 min)⇒Tokyo Station＝(95 min)⇒Sendai Station

### ▶①仙台城

伊達政宗が青葉山に築城した仙台城。敷地にある展望台からは、仙台市を一望することができます。

### ▶①Sendai Castle

The Sendai Castle was built by Masamune Date in Mt. Aoba. From the sightseeing tower in the area, you can see the entire town of Sendai.

### ▶②大崎八幡宮

「厄除け・除災招福や必勝・安産」の神として崇められた大崎八幡宮。伊達政宗の寄進により建立されたこの神社の社殿（本殿・石の間・拝殿）は、国宝に指定されています。

### ▶②Osakihachimangu Shrine

The Osakihachimangu Shrine is known for revering the god of warding off evils and easy delivery. The shrines were built by the contribution of Masamune Date and are registered as the national treasures.

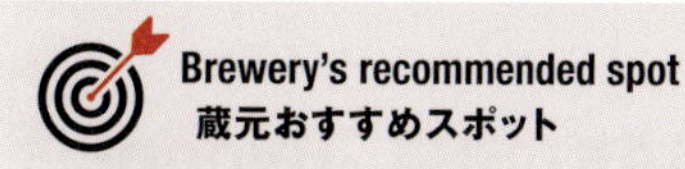


## 瑞鳳殿

仙台市青葉区霊屋下23-2
23-2, Otamayashita, Aoba-ku Sendai-shi

初代仙台藩主伊達政宗公が眠る霊廟所。仙台の観光名所。

### Zuihoden Mausoleum

It is where the first feudal lord of Sendai han, Masamune Date sleeps. It is a tourist spot in Sendai.

## 広瀬川

### Hirose River

A river running through Sendai city. You can see it from the city and is the symbol of Sendai.

仙台市を貫流する川。市街地からは自然景観が楽しむことができ、仙台市のシンボルとなっている。

## 旅酒が見つかる場所 Shop

**鷹泉閣 岩松旅館**
仙台市青葉区作並温泉元湯
022-395-2211

**仙台ヒルズホテル＆ゴルフ倶楽部**
仙台市泉区実沢中山南25-5
022-719-8711

※最新の販売店情報はホームページでご確認ください。
Please check our website for the latest retail shop information.

**◆他にも素敵な場所があります。あなただけの旅の宝を探してみましょう!**
**There are other wonderful places to visit. Let's take a trip to find a treasure of your own!**

48

## HIMEJI・KINOSAKI SPA -Hyogo-

# 姫路・城崎温泉 （兵庫県）

日本100名城に登録されている姫路城を有する姫路エリアと平安時代からの長い歴史をもつ城崎温泉。兵庫県内では外せない異なる魅力を持つ観光スポット。全国のみならず、海外からも多くの旅行客が訪れるこの2つのスポットをぜひお楽しみください。

The Himeji Castle, which is one of the 100 famous castles in Japan, and Kinosaki Onsen, which has a long history from Heian period. These are spots where you would want to visit when you visit Hyogo prefecture. Not only Japan, but also many foreign visitors come to visit these two spots. Please enjoy when you have a chance to visit.

## 旅 TABI-SAKE 酒

HIMEJI ・ KINOSAKI SPA

天然水を使用し米と米麹のみで仕上げた純米酒です。ソフトな口あたりですっきり感があり淡々と飲み飽きのこない酒質に仕上げています。

Using the natural water and making the sake with just rice and malted rice. The texture is very soft and is refreshing so you won't get tired of the taste.

**醸造元**：名城酒造株式会社
（1864年創業／兵庫県）
**種類**：日本酒 純米酒
**度数**：14.8度
**精米歩合**：70%

**Brewery** : Meijo Sake Brewing Co., Ltd.（Since 1864／Hyogo）
**Kind** : Junmai-shu
**Alcohol Content** : 14.8%
**Polishing Ratio** : 70%

甘辛度
Sweet-dryness
甘 Sweet — 辛 Dry

## 主な行き方 Major Directions

東京駅＝（新幹線180分）⇒姫路駅＝（特急85分）⇒城崎温泉駅
Tokyo Station＝(Shinkansen180 min)⇒Himeji Station＝(express85 min)⇒Kinosakionsen Station

## 旅の宝 Treasure Hunt

### ▶①姫路城

1346年に築城され、長い歴史を持つ姫路城。国宝、世界文化遺産など、さまざまな指定文化財に登録されている日本の史跡として代表的なもののひとつです。平成の大改修も終え、国内外から多くの観光客が訪れる人気スポット。

### ▶①Himeji Castle

The Himeji Castle was built in 1346 and has a long history. It is registered as the World Heritage Site, the national treasure and many other cultural assets. It has finished its fundamental reconstruction, and many foreign and domestic tourists are visiting.

### ▶②城崎温泉

開湯1300年の歴史を持つ兵庫県随一の名湯地。石造りの太鼓橋や川沿いの柳並木が情緒ある風景で、近年では外国人にも人気です。城崎名物といえば「外湯めぐり」といわれ、温泉街には7軒もの外湯が点在しています。

### ▶②Kinosaki Onsen

It has a history of 1300 years since it opened. The scenery is very Japanese like with the stone bridges and the cherry blossoms by the river. The most attractive part of the Kinosaki Onsen is the “Sotoyu Meguri” where you can enjoy 7 different hot springs.

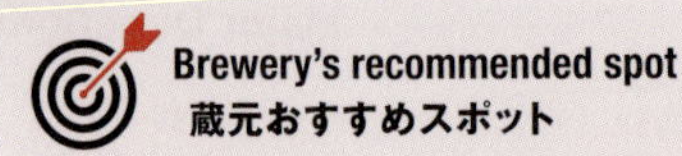


## 竹田城

朝来市和田山町竹田古城山169

169, Kojozan, Takeda, Wadayama-cho, Asago-shi

現存する山城として日本屈指の規模を持つ竹田城。雲海に包まれた姿は、天空に浮かぶ城として、その幻想的な景観が人気のスポット。天空の城という異名を持つこの城の景観は、みる人を圧倒する。

## Takeda Castle

The Takeda Castle is one of the largest Yamajiro (mountain castle) in Japan. The castle in full of clouds looks like the castle floating in the air, and the view is very famous. It is also called as the Castle in the sky, and the scenery is overwhelming.

世界遺産・姫路城を借景にした本格的な日本庭園で面積は約一万坪。
池泉回遊式の御屋敷の庭など9つの庭園群で構成され江戸の情緒を醸し出す。

## 好古園

姫路市本町68

68, Hommachi, Himeji-shi

## Kokoen Garden

It is a world heritage which is a Japanese garden, 10,000 tsubo big, and has the Himeji Castle at the back. There are nine different types of garden and one of them is a style of garden that features a path around a pond.

### 旅酒が見つかる場所 Shop

- **山陽百貨店**
  姫路市南町1
  079-223-1231
- **日本酒bar 試**
  姫路市南駅前町125
  ピエラ姫路
  079-280-6172
- **岡本酒店**
  豊岡市城崎町湯島118
  0796-32-2138

※最新の販売店情報はホームページでご確認ください。
Please check our website for the latest retail shop information.

## BOSO -Chiba-

# 房総 (千葉県)

温暖な気候に海と山、首都圏からのアクセスも便利な観光地、房総。海の幸、山の幸を味わうグルメ旅行にも、動物と触れ合うテーマパーク巡りにも、大自然を満喫するハイキングにもおすすめです。

The Boso area is a touring spot easily accessible from the city and is rich in nature. There are many ways to enjoy in the area such as eating seafood and vegetables, visiting theme parks where you can play with animals, and hiking the great nature of Boso.

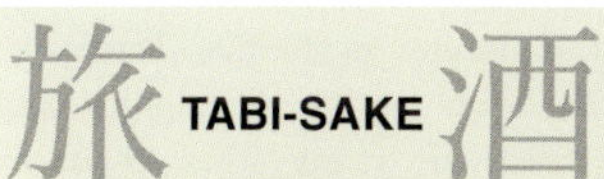

## 旅 TABI-SAKE 酒

**BOSO**

山廃の旨みと酸を感じられる辛口の純米酒。

Using natural yeast, the Junmai-shu is dry and sour.

**醸造元**：岩瀬酒造株式会社
（1723年創業／千葉県）
**種類**：日本酒 純米酒
**度数**：15度
**精米歩合**：70%

**Brewery** : Iwase Syuzo Corp.（Since 1723／Chiba）
**Kind** : Junmai-shu
**Alcohol Content** : 15%
**Polishing Ratio** : 70%

**甘辛度**
Sweet-dryness

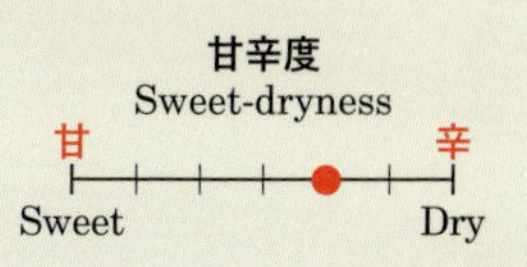

## 主な行き方　Major Directions

羽田空港＝(20分)⇒浜松町駅＝(40分)⇒東京駅＝(45分)⇒千葉駅＝(6分)⇒蘇我＝(特急45分)⇒御宿

Haneda Airport＝(20 min)⇒Hamamatsucho Station＝(40 min)⇒Tokyo Station＝(45 min)⇒Chiba Station＝(6 min)⇒Soga Station＝(express 45 min)⇒Onjuku Station

### ▶①鋸山

日本最大の大仏がある鋸山。断崖が垂直に切り立つ「地獄のぞき」は、崖の上から房総半島を一望できる絶景ポイント。

### ▶①Mt. Nokogiri

There is the largest statue of Buddha at the Mt. Nokogiri. At the point where the cliff is cut vertically, “peeking hell,” you can see the whole view of the Boso region.

### ▶②鴨川シーワールド

房総半島にある大規模な総合海洋レジャーセンター。イルカやアシカのショーをはじめ、さまざまなイベントを行っています。

### ▶②Kamogawa Seaworld

The Kamogawa Seaworld is a huge aquarium located in the Boso Peninsula. Starting with shows of dolphins and seals, there are many other events as well.

**Brewery's recommended spot**
**蔵元おすすめスポット**

## 月の沙漠記念館
## ラクダ像

東隅郡御宿町六軒町505-1

505-1, Rokken-machi, Onjuku-machi, Isumi-gun

御宿をこよなく愛した詩人加藤まさをの作品や資料の展示・公開をはじめ御宿にゆかりのある文人や画家たちの紹介等、御宿の再発見と新しい文化の創造を目指して建てられた夢とロマンあふれる記念館。

## Tsuki no Sabaku Memorial Museum and Camel Statue

It is a memorial museum of Masao Kato, who was a poet and loved Onjuku, and his arts and documents are displayed. He also introduced people who had relation with Onjuku; it is a museum full with romance and dreams.

## メキシコ塔

夷隅郡御宿町岩和田1016

1016, Iwawada, Onjuku-machi, Isumi-gun

岩和田海岸（現田尻浜）で沈没したイスパニア（スペイン）船・サンフランシスコ号を岩和田の人達は力を合わせて救助したことから、日本とメキシコとスペインの三国の交通が始まった。それを記念して建てられたのがメキシコ塔。

## Mexico Tower

The people of Iwawada rescued people who were riding the Spain ship that sank in the Iwata Coast, and that was the beginning of Japan, Mexico, and Spain started trading. It is the memorial tower of that event.

**旅酒が見つかる場所** **Shop**

- **道の駅 和田浦 WA・O！**
  南房総市和田町仁我浦243
  0470-47-3100
- **セブン-イレブン 上総御宿店**
  夷隅郡御宿町須賀605-3
  0470-68-2828

※最新の販売店情報はホームページでご確認ください。
Please check our website for the latest retail shop information.

50

## SAPPORO・OTARU・HAKODATE -Hokkaido-

# 札幌・小樽・函館 （北海道）

北海道の代表的な町、札幌・小樽・函館。それぞれに異なる魅力がある町であり、旅行者が巡る代表的な観光ルートにもなります。道庁所在地でもある札幌は、都市の魅力と大自然が見事に調和したエリアです。運河や石造りの倉庫の景観が美しい小樽、函館朝市等グルメも魅力的な函館も一度は訪れたいスポットです。

Sapporo, Otaru, and Hakodate are the famous cities in Hokkaido. Each city is different from one another and they are all popular from the tourists.
Sapporo, which is the prefectural capital of Hokkaido, is an area where the charm of the city and nature are harmonized perfectly. Otaru has beautiful scenery of canals and stone warehouses. In Hakodate, the gourmet food in the morning market is attractive. Those cities are definitely worth visiting.

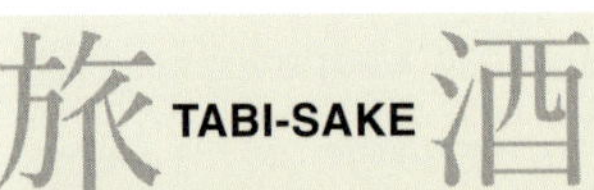


**SAPPORO · OTARU · HAKODATE**

北海道の大地で育った酒造好適米と水で仕込んだ“どさんこ”の日本酒。米の旨みを十分に引き出した、柔らかで口あたりの良い純米酒です。

The Special Junmai-shu using the rice grown in Hokkaido and the “Dosanko.” Its texture is very soft and is full of rice umami flavor.

**醸造元**：男山株式会社
（1887年創業／北海道）
**種類**：日本酒 特別純米酒
**度数**：15度
**精米歩合**：55%

**Brewery** : Otokoyama Co., Ltd.
(Since 1887／Hokkaido)
**Kind** : Special Junmai-shu
**Alcohol Content** : 15%
**Polishing Ratio** : 55%

**甘辛度**
Sweet-dryness
甘 ← → 辛
Sweet ― Dry

## 主な行き方 Major Directions

羽田空港＝(90分)⇒新千歳空港＝(40分)⇒札幌駅
Haneda Airport＝(90 min)⇒New Chitose Airport＝(40 min)⇒Sapporo Station

羽田空港＝(90分)⇒新千歳空港＝(80分)⇒小樽駅
Haneda Airport＝(80 min)⇒New Chitose Airport＝(80 min)⇒Otaru Station

羽田空港＝(80分)⇒函館空港／バス＝(20分)⇒函館駅
Haneda Airport＝(80 min)⇒Hakodate Airport／Bus＝(20 min)⇒Hakodate Station

## 旅の宝 Treasure Hunt

### ▶①小樽運河

運河沿いにある石造倉庫群が完成された大正時代の状態のまま残っている、小樽を代表する観光スポットのひとつ、小樽運河。ガラス・オルゴール工房や飲食店、海産物店、菓子店、土産物店など、さまざまなお店が入っています。

### ▶①Otaru Canal

Along the canal, there are stone-built warehouses from the Taisho period which are famous spots to visit. In the warehouse, there are many stores such as glass music box store, restaurants, snacks, and souvenir stores.

### ▶②箱館山

函館の町を一望できる箱館山。ミシュランにも3つ星を受けた日本の代表的夜景のひとつとして人気のエリア。世界でも屈指の人気と知名度を誇る函館の夜景は必見です。

### ▶②Mt. Hakodate

The Mt. Hakodate is a place where you can see the entire city of Hakodate. The night view from this mountain has received three stars from the Michelin. You should see the night view of Hakodate when you have a chance to visit.

## Brewery's recommended spot
## 蔵元おすすめスポット

### 定山渓温泉

札幌市南区定山渓温泉
Jozankeionsen, Minami-ku, Sapporo-shi

河童伝説が残り河童をシンボルにしている。札幌の奥座敷として人気が高く、札幌から日帰りで温泉を楽しめる。

### Jozankei Onsen

The symbol is a Kappa since it has legends of it. You can enjoy the hot spring from Sapporo as an one day trip and is very popular in Sapporo.

### 幌美峠

札幌市中央区盤渓471-110
471-110, Bankei, Chuo-ku, Sapporo-shi

### Horomi Pass

When you want to see lavenders, but you don't have time for Furano, you should go to the Horomi Pass to see the lavenders.

富良野まで足を延ばす時間がない時は、札幌の中央区にある幌見峠にあるラベンダー園を訪れてみては。

## 旅酒が見つかる場所 Shop

- **北海道ブランドショップ**
  小樽市築港11
  ウィングベイ小樽
  イオン小樽店 1階
  0134-21-6121
- **柿崎商店**
  余市郡余市町黒川町7
  0135-22-3354
- **エルラプラザ**
  余市郡余市町黒川町5-43
  0135-22-1515
- **酒舗 稲村屋**
  北斗市市渡1-1-7
  北斗市観光交流センター別館
  「ほっくる」内
  0138-83-8565
- **男山酒造り資料館**
  旭川市永山2-7-1-33
  0166-47-7080
- **（株）アートクリエイト**
  小樽市港町4-5
  0134-32-0111
- **北緯43度**
  岩内郡岩内町万代47-9
  0135-62-8343
- **ビックリッキー 黒松内店**
  寿都郡黒松内町黒松内197
  0136-72-3011
- **前田商店**
  虻田郡ニセコ町字本通81-3
  0136-44-2433
- **矢田商店**
  岩内郡岩内町清住7-5
  0135-62-1104
- **函館湯の川温泉 湯元 啄木亭 売店**
  函館市湯川町1-18-15
  0138-59-5355
- **苫小牧東港周文 フェリーターミナル**
  勇払郡厚真町浜厚真17
  0145-28-2800
- **（有）山本酒店**
  網走市大曲1-14-19
  オホーツクバザール内

※最新の販売店情報はホームページでご確認ください。
Please check our website for the latest retail shop information.

# SHIRETOKO・AKAN -Hokkaido-

# 知床・阿寒（北海道）

51

世界遺産登録地域の知床と、マリモで有名な阿寒湖。北海道の自然遺産を満喫することができるこの地域は道東観光の代表ルートです。冬に見られる流氷の絶景や、夏に楽しむ北海道の大自然の中のトレッキングなど、自然を体感するには最適な観光地です。

Shiretoko area is registered as the World's Heritage Site and Lake Akan is famous for marimos. This area, where you can feel the nature, is a famous touring spot in Eastern Hokkaido.
It is a superb sightseeing spot to experience nature such as the spectacular views of drift ice in winter and trekking in the great nature of Hokkaido in summer.

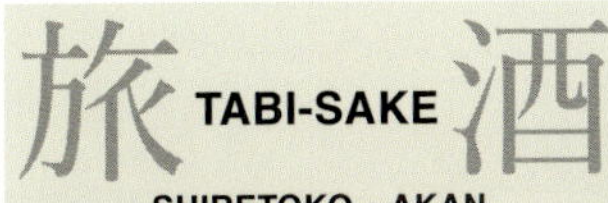


SHIRETOKO · AKAN

北海道の大地で育った酒造好適米と水で仕込んだ"どさんこ"の日本酒。米の旨みを十分に引き出した、スッキリとした飲み口の本醸造酒です。

The Special Honjozo-shu using the rice grown in Hokkaido and the "Dosanko." Its texture is very refreshing.

**醸造元**：男山株式会社
（1887年創業／北海道）
**種類**：日本酒 特別本醸造酒
**度数**：15度
**精米歩合**：60%

**Brewery** : Otokoyama Co., Ltd.
(Since 1887／Hokkaido)
**Kind** : Special Honjozo-shu
**Alcohol Content** : 15%
**Polishing Ratio** : 60%

**甘辛度**
Sweet-dryness

甘 ─┼─┼─┼─┼─●─┼─ 辛
Sweet　　　　　　　Dry

## 主な行き方 Major Directions

羽田空港＝(105分)⇒女満別空港／バス＝(30分)⇒網走駅＝(50分)⇒知床斜里駅

Haneda Airport＝(105 min)⇒Memanbetsu Airport／Bus＝(30 min)⇒Abashiri Station＝(50 min)⇒Shiretoko Shari Station

羽田空港＝(100分)⇒釧路空港／バス＝(80分)⇒阿寒湖

Haneda Airport＝(100 min)⇒Kushiro Airport／Bus＝(80 min)⇒Lake Akan

## 旅の宝 Treasure Hunt

### ▶①知床五湖

知床八景のひとつに数えられる観光地、知床五湖。原生林に囲まれてたたずむ幻想的な5つの湖の景色は必見です。

### ▶②阿寒湖

阿寒摩周国立公園にある、北海道で5番目に大きい淡水湖。日本で数少ないマリモが生息する湖です。北海道最大のアイヌコタン（アイヌの集落）もあり、多くの観光客が訪れています。

### ▶①Shiretoko Goko Lakes

Shiretoko Goko Lakes is one of the eight famous tour spots. The five lakes surrounded by virgin forests is a must-see fantastic spot.

### ▶②Lake Akan

The Lake Akan is the fifth largest lake in Hokkaido located in the Akan-Mashu National Park. There are marimos living in this lake. There is also the largest Ainu village, and may tourists visit there.

### Brewery's recommended spot 蔵元おすすめスポット

オシンコシンの滝

斜里町ウトロ

Utoro, Shari-cho

国道334号沿いにある知床半島最大の滝。途中で二筋に分かれる滝であるため「双美の滝」の名もある。

Oshinkosin Waterfall

The largest waterfall in Shiretoko, where it is along the national road 334. It is also called the soubi no taki since it separates into two streams.

## 神の子池

斜里郡清里町清泉
Kiyoizumi, Kiyosato-cho, Shari-gun

原生林に囲まれたコバルト色の神秘的な池。池の底から摩周湖の伏流水と言われている湧水が大量に湧き出している。

## Kaminoko Pond

A pond with cobalt blue color covered by trees. From the bottom of the lake, there are water springs that stream to the Lake Mashu.

## 天に続く道

斜里群斜里町峰浜～大栄地区（国道334号／244号）
Minehama, Shari-cho, Shari-gun ~ Taiei area (Route 334/244)

全長約18kmの直線道路。
まっすぐな道が、高低差により道の先が天まで続いているように見えることからこの名前がついた。

## Road to Heaven

It is a straight street that is 18 km long. It is called the Ten no Michi because it looks like as if the street is going to the sky.

### 旅酒が見つかる場所 Shop

- **男山酒造り資料館**
  旭川市永山2-7-1-33
  0166-47-7080
- **（株）アートクリエイト**
  小樽市港町4-5
  0134-32-0111
- **北緯43度**
  岩内郡岩内町万代47-9
  0135-62-8343
- **ビックリッキー 黒松内店**
  寿都郡黒松内町黒松内197
  0136-72-3011
- **前田商店**
  虻田郡ニセコ町字本通81-3
  0136-44-2433
- **矢田商店**
  岩内郡岩内町清住7-5
  0135-62-1104
- **函館湯の川温泉 湯元 啄木亭 売店**
  函館市湯川町1-18-15
  0138-59-5355
- **苫小牧東港周文 フェリーターミナル**
  勇払郡厚真町浜厚真17
  0145-28-2800
- **道の駅ウトロ シリエトク ユートピア知床**
  斜里郡斜里町ウトロ西186-8
  0152-24-5160
- **グランマルシェ 鶴雅店売店**
  釧路市阿寒町阿寒湖温泉4-6-10
  0154-67-2531
- **（有）山本酒店**
  網走市大曲1-14-19
  オホーツクバザール内

※最新の販売店情報はホームページでご確認ください。
Please check our website for the latest retail shop information.

52

# KORIYAMA -Fukushima-

# 郡山 （福島県）

郡山は市街地と自然が共存する街。東京から東北新幹線で1時間20分程で来られる東北第二の商業都市で、ビジネスや福島観光の拠点としても有名な都市です。

Koriyama is a town where its urban area and the surrounding nature coexist. It is the second largest commercial city in Tohoku, about an hour and 20 minutes from Tokyo by the Tohoku Shinkansen. It is famous as a base for business and sightseeing in Fukushima prefecture.

## 旅 TABI-SAKE 酒

KORIYAMA

厳選したモルトウイスキー原酒にグレーンウイスキーをブレンドし香り良くマイルドで口あたりが良い味わい。

Made from a blend of carefully selected malt whiskies and grain whiskies, it offers a mild and smooth taste with a pleasant scent.

**醸造元**：笹の川酒造株式会社
（1765年創業／福島県）
**種類**：ウイスキー
**度数**：40度

**Brewery** : SASANOKAWA SHUZO Co., Ltd.（Since 1765／Fukushima）
**Kind** : Whiskey
**Alcohol Content** : 40%

## 主な行き方 — Major Directions

羽田空港＝(20分)⇒浜松町駅＝(5分)⇒東京駅＝(80分)⇒郡山駅

Haneda Airport＝(15 min)⇒Hamamatsucho Station＝(5 min)⇒Tokyo Station＝(80 min)⇒Koriyama Station

## 旅の宝 Treasure Hunt

### ▶①郡山布引風の高原

猪苗代湖南側にある布引高原は標高1000mに位置し、眺望がすばらしいスポットです。風力発電用の風車が並び、高原一面に広がった四季折々の花々と一緒に写真に納める観光客が多く訪れます。

### ▶①Koriyama Nunobiki Wind of Plateau

The Nunobiki Plateau, located to the south of Lake Inawashiro, is 1,000 meters above sea level and has a great view. With wind turbines for wind power generation lined up, the plateau has a lot of visitors who enjoy viewing seasonal flowers and taking pictures of its beautiful scenery.

### ▶②猪苗代湖

郡山のシンボルでもある猪苗代湖は、全国で4番目の広さの大きな湖です。「湖南七浜」と呼ばれる7つの浜は、夏になるとマリンスポーツを楽しむ方々で賑わいます。冬には「しぶき氷」が見られることでも有名。

### ▶②Lake Inawashiro

Lake Inawashiro, the symbol of Koriyama, is the fourth largest lake in Japan. The seven beaches called "Konan Nanahama" become lively with people enjoying marine sports in summer. It is also famous for a natural phenomenon called "Shibuki-gori" (literally "splashed ice") in winter.

Brewery's recommended spot
蔵元おすすめスポット

## 開成山公園

郡山市開成1-5
1-5, Kaisei, Koriyama-shi

五十鈴湖を中心に整備された公園。800本以上のバラが咲き誇るバラ園は見頃の6月、10月に一般公開され多くの方が訪れる。また、公園内には1300本の桜の木があり、隣の開成山大神宮の桜とあわせ、郡山の桜の名所となっている。

## Kaiseizan Park

It is a park established around Lake Isuzu. The rose garden which has more than 800 roses is open to the public in June and October when they are in full bloom, and many visitors enjoy viewing them. There are 1,300 cherry trees in the park, and along with the neighboring Kaiseizandaijingu Shirine, the park is known as a beautiful cherry-blossom spot in Koriyama.

**旅酒が見つかる場所** — **Shop**

現在販売準備中です。
Currently in preparation.

※最新の販売店情報はホームページでご確認ください。
Please check our website for the latest retail shop information.

◆**他にも素敵な場所があります。あなただけの旅の宝を探してみましょう!**
**There are other wonderful places to visit. Let's take a trip to find a treasure of your own!**